Games, Logic, and Constructive Sets

CSLI Lecture Notes
Number 161

Games, Logic, and Constructive Sets

edited by
Grigori Mints
Reinhard Muskens

CSLI PUBLICATIONS
Center for the Study of Language and Information
Stanford, California

Copyright © 2003
CSLI Publications
Center for the Study of Language and Information
Leland Stanford Junior University
Printed in the United States
07 06 05 04 03   5 4 3 2 1
Library of Congress Cataloging-in-Publication Data
Games, logic, and constructive sets /
edited by Grigori Mints and Reinhard Muskens.
p.   cm. (CSLI lectures notes ; no. 161)
Includes bibliographical references and index.
ISBN 1-57586-449-5 (alk. paper)
ISBN 1-57586-450-9 (pbk. : alk. paper)
1. Logic, Symbolic and mathematical. 2. Game theory.
3. Constructibility (Set theory)
I. Mints, G. E. II. Muskens, Reinhard, 1953-
III. Title. IV. Series.
BC135.G26   2003
160–dc22   2003015819
CIP

∞ The acid-free paper used in this book meets the minimum requirements of the American National Standard for Information Sciences—Permanence of Paper for Printed Library Materials, ANSI Z39.48-1984.

CSLI was founded in 1983 by researchers from Stanford University, SRI International, and Xerox PARC to further the research and development of integrated theories of language, information, and computation. CSLI headquarters and CSLI Publications are located on the campus of Stanford University.

CSLI Publications reports new developments in the study of language, information, and computation. In addition to lecture notes, our publications include monographs, working papers, revised dissertations, and conference proceedings. Our aim is to make new results, ideas, and approaches available as quickly as possible. Please visit our web site at
http://cslipublications.stanford.edu/
for comments on this and other titles, as well as for changes and corrections by the author and publisher.

Contents

I Logic and Games   1

**Logic and Game Theory: Close Encounters
of the Third Kind**
Johan van Benthem   3

- 1   Encounters of three kinds 3
- 2   Games from a logical viewpoint 4
  - 2.1   Pictures 4
  - 2.2   Games as models for logical languages 5
  - 2.3   Intensional logic as usual 7
  - 2.4   Major issues in process theories 7
- 3   Levels of representation 8
  - 3.1   Games: from actions to outcomes 8
  - 3.2   Actions in modal and dynamic logic 9
  - 3.3   Powers and a forcing language 11
  - 3.4   Intermediate game levels 12
- 4   Game operations and game algebra 13
  - 4.1   Game operations 13
  - 4.2   Game algebra 14
  - 4.3   Languages 14
- 5   From finite to infinite games 15
- 6   Coping with imperfect information 16
  - 6.1   From perfect to imperfect information 16
  - 6.2   Uniform outcome equivalence and a forcing modality 17
  - 6.3   Actions and information: dynamic-epistemic logic 18
- 7   Preferences and rational behavior 19
  - 7.1   Game values, equilibrium and backward induction 19
  - 7.2   Logic issues revisited 20
- 8   Conclusion 21

v

Informationally Independent Connectives
Gabriel Sandu and Ahti Pietarinen   23
- 1   Sentential logic and games 23
  - 1.1   Extensive games of perfect information 24
  - 1.2   Extensive games of imperfect information 25
  - 1.3   Extensive forms as Kripke frames 30
- 2   Informationally independent connectives 33
  - 2.1   Games of perfect information 33
  - 2.2   Semantical games of imperfect information . . . 35
  - 2.3   Informationally independent connectives and knowledge of the game 39

Descriptions of Game States
Hans van Ditmarsch, Wiebe van der Hoek, Barteld Kooi   43
- 1   Introduction 43
- 2   Epistemic logic 44
- 3   Description of Hexa 46
  - 3.1   Derived characteristics of Hexa 48
- 4   Description of initial game states 49
  - 4.1   Derived characteristics: factual knowledge 51
  - 4.2   Derived characteristics: private knowledge 52
  - 4.3   Derived characteristics: private ignorance 52
  - 4.4   Derived characteristics: seedontknow 54
- 5   Description of the pre-initial state 54
- 6   Further observations 55
- 7   Conclusion 57

II   Classical Logic   59

Resource Consciousness in Classical Logic
Andreas Blass   61
- 1   Introduction 61
- 2   An Example 62
- 3   Herbrand's Theorem 63
- 4   Simple Herbrand Validity 67
- 5   Universal Simple Herbrand Validity 68
- 6   Modus Ponens 72
- 7   Connection with Affine Logic 73

Quick Cut-Elimination for Monotone Cuts
Grigori Mints   75
1. Introduction 75
2. System LK 77
3. Cut-Elimination 78

III   Constructive Set Theory   85

The Anti-Foundation Axiom in Constructive Set Theories
Michael Rathjen   87
1. Introduction 87
2. The anti-foundation axiom 88
3. **AFA** in constructive set theory 89
    3.1   The theory **CZFA** 89
    3.2   Interpreting **AFA** in Martin-Löf type theory . . 92
    3.3   Upper bounds 96
    3.4   Lower bounds 98
4. Anti-foundation with inaccessible sets 100

On Non-wellfounded Constructive Set Theory: Construction of Non-wellfounded Sets in Explicit Mathematics
Sergei Tupailo   109
1. Constructive Set Theory with Natural Numbers 109
2. Explicit Mathematics: a reminder 111
3. Realization of **NCZF**$^-$ into **EETJ** 113

Index   **127**

Preface

The papers collected in this volume are mostly contributions to the Ninth CSLI Workshop on Logic, Language and Computation, held at Stanford University at the end of May 2000. The workshop was organized by David Beaver, Johan van Benthem, Rob van Glabbeek, David Israel, Grigori Mints, and Patrick Scotto di Luzio. It was sponsored by the Center for the Study of Language and Information (CSLI), the Departments of Linguistics and Philosophy, the School of Humanities and Sciences, Professor Vaughan Pratt (all at Stanford University) and by Johan van Benthem's Spinoza Project 'Logic in Action', the Netherlands. Special thanks go to all those involved in the anonymous reviewing process.

Grigori Mints
Reinhard Muskens

Contributors

JOHAN VAN BENTHEM is Professor of Computer Science, University of Amsterdam, and Professor of Philosophy, Stanford University. His main interests are in the logical analysis of action, information and communication.

ANDREAS BLASS is Professor of Mathematics at the University of Michigan. His interests include set theory, category theory (particularly topos theory), theoretical computer science (particularly computational complexity), linear logic, and combinatorics.

HANS VAN DITMARSCH is a Lecturer at the Computer Science Department of the University of Otago. He has worked as a lecturer at CS and AI departments at various Dutch universities, mainly at the University of Groningen and at the Open University of the Netherlands. He wrote a PhD thesis on logic and games.

WIEBE VAN DER HOEK is Associate Professor at the Institute of Information and Computing Sciences of the University of Utrecht and Professor at the Computer Science Department of the University of Liverpool. His current interests are logics for AI, belief revision and game theory.

BARTELD KOOI is currently working on a PhD project at the Department of Mathematics and Computing Science at the University of Groningen under the supervision of Gerard Renardel and Rineke Verbrugge. His scientific interests are epistemic logic, dynamic logic, probability logic, and game theory.

GRIGORI MINTS is Professor of Philosophy at Stanford University. He is interested in foundations of mathematics, proof-theoretic methods and their applications to philosophical logic and computer science.

REINHARD MUSKENS is Associate Professor at the Linguistics Department of Tilburg University. His main research interest is in applications of logic to linguistic theory.

AHTI PIETARINEN is at the University of Helsinki, Department of Philosophy. His research interests are in the area of semantic games in logic and language.

MICHAEL RATHJEN is Professor of Mathematics at the University of Leeds. His interests are in proof theory, ordinal analysis, constructive set theory, explicit mathematics, and type theory.

GABRIEL SANDU is currently Professor at the Department of Philosophy, University of Helsinki. His first degree was in Economics at the Academy of Economic Sciences of Romania. His main areas of interest are semantics for natural languages, games, and formal theories of truth.

SERGEI TUPAILO is a researcher at the University of Bern, Switzerland. His interests are in proof theory and constructive mathematics, in particular Hilbert's epsilon substitution method, ordinal analysis, explicit mathematics, and frameworks of constructivism.

Part I
Logic and Games

Logic and Game Theory: Close Encounters of the Third Kind

Johan van Benthem

> This paper is an exercise at the interface of logic and game theory. We show how games may be analyzed in the style of logical process theories, starting from the pure action case with perfect information, and then including more realistic features like imperfect information and preferences.

1 Encounters of three kinds

Games are a species of processes with intricate dynamic interactions between players about which we have vivid first-hand intuitions. Historically, this richness has led to several mathematizations, starting with the probabilistic one of the 17th century, emphasizing odds and outcomes in betting games. In the 20th century, economic game theory developed a powerful account of strategic games of various kinds, and equilibria between strategies. But there is still a third level of interest: the fine-structure of players' actions, deliberations, and decisions as they move along — and here contemporary logic is beginning to meet game theory. There are several strands to this contact. Logicians have long employed *logic games* for analyzing argumentation, semantic evaluation, model comparison, or model construction. Winning strategies in these games capture central logical notions such as proof, truth, or similarity. These are very specialized games from a game-theoretical viewpoint, lacking finer utilities for players, and assuming perfect information about any play. But there is also logical structure to the general games in the game-theoretic literature. The first, and most

---

*Address: Institute for Logic Language and Computation (ILLC), University of Amsterdam, Plantage Muidergracht 24, 1018 TV Amsterdam, The Netherlands. April–June: Center for the Study of Language and Information (CSLI), Stanford University, Ventura Hall, Stanford 94305-4115. Email: johan@science.uva.nl. Home page: http://staff.science.uva.nl/~johan/.

Games, Logic, and Constructive Sets
Grigori Mints and Reinhard Muskens (eds.)
Copyright ©2003, CSLI Publications

established trend at the interface comes from *philosophical logic*. Reasoning by players in the course of a game, or by an observer making sense of the game, naturally involves notions like knowledge, belief, and belief revision. This is the area of epistemic logic, conditional logic, and belief revision theory — and there has even been an independent rediscovery of epistemic logic inside the game theory of the 1970s. Much of the interest here focuses on a better understanding of *rationality* of actors in a complex setting. The second, less established trend comes from *computer science*. It looks at games as distributed processes involving intelligent agents, an interactive setting that is typical for modern internet-based computational tasks, from information processing to electronic commerce. Tools for analyzing game structure in this sense include dynamic logic, temporal logic, and theories of concurrency like process algebra. One major concern here is the shifting balance of *expressive power* and *algorithmic complexity* in complex interactive processes.

This paper is a short introduction to these interfaces in the process perspective with a sprinkling of epistemic concerns. We refer to other papers for proofs of results: our interest here is the over-all setting. Moreover, our discussion presupposes some acquaintance with the essentials of modal logic and game theory (cf. Blackburn et al. (2001); Osborne and Rubinstein (1994)).

2 Games from a logical viewpoint

2.1 Pictures

To most people, 'game theory' conjures up pictures like the 2x2 matrices for Prisoner's Dilemma — or less worn-out examples like the 'Stag Hunt', where two players must choose one of two available strategies independently:

	hunt stag	*hunt rabbit*
hunt stag	3,3	0,1
hunt rabbit	1,0	1,1

The story is this. The game has two 'equilibria': one where both players hunt stag (an arduous enterprise requiring cooperation, yielding a high pay-off for each), and one where both hunt rabbit (individual, less arduous, but also less rewarding). Choosing different strategies is not a stable form of joint behaviour: a stag hunter confronted with a rabbit hunter is better off switching to rabbit hunting. More generally, a pair of strategies is in *Nash equilibrium* if neither player can improve her outcome by deviating when the other player sticks to his strategy.

Other typical game-theoretic pictures are 'Centipedes', such as the game depicted here

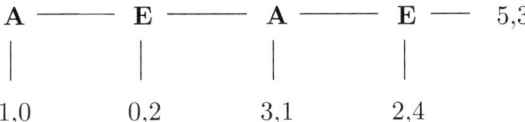

with values for outcomes indicated in the displayed pairs of (**A**-value, **E**-value). Here the story is as follows. Players can move 'down' or 'right'. Arguing backward from the right, at each state of the game, the active player is best off going down, leading to a prediction that the game will stop after **A**'s first move. This is surprising, because the far-right outcome (5, 3) seems certainly better for both. The Centipede has generated much controversy about 'rational behaviour' — but we just display it for its structure. As opposed to the Stag Hunt, a global *strategic game*, it is an extensive game displaying all local moves. Moreover, it has *perfect information*: players always know what the other has done so far. In the Stag Hunt, however, players choose their actions in ignorance of what the other did: a form of *imperfect information*. Now, what does all this have to do with logic?

2.2 Games as models for logical languages

Actions in modal and dynamic logic The most obvious foothold for logical analysis is in extensive games. Consider the following 'game form' for two players **A**, **E**, involving states (nodes in the tree), moves (labeled arrows), runs (maximal sequences of successive moves), and outcome states for the runs:

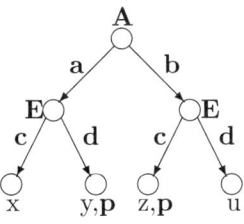

Figure 1

Just as it stands, this tree is a model for a *modal logic* of actions (reachability via players' moves) with further special structure encoded in local properties indicating turns at intermediate nodes, and other relevant properties **p** at nodes. One reads this picture with a global dynamic intuition. The game involves players' *strategies*, assigning an available move to every node where it is their turn. Playing a strategy guarantees an outcome in the set of end states arising from every possible counter-play by the other player. This is the associated *power* for forcing the game into a specified set of outcomes. In this particular game,

A has 2 strategies, with powers {x, y}, {z, u}
E has 4 strategies, with powers {x, z}, {x, u}, {y, z}, {y, u}

Strategic powers can be described in a *dynamic logic* allowing complex actions. E.g., with ∪ standing for choice between moves, the following box-diamond formula says that player **E** has a strategy forcing a set of outcomes satisfying **p**:

$$[a \cup b]\langle c \cup d\rangle p$$

Preference logics Now consider the preceding game with *players' preferences* added, in the form of utility values for both players at outcome states:

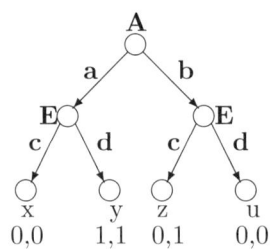

| values for **A**, **E** | x
0,0 | y
1,1 | z
0,1 | u
0,0 |

Figure 2

A utility-maximizing agent **E** will play strategy ⟨left:d, right:c⟩, and taking this into account, a rational agent **A** plays move a. Richer models like this interpret *preference logics* of various sorts, such as conditional logic, or logics of belief. E.g., we may naturally say that players believe the game will end in y — at least assuming that everyone plays rationally, so that 'preference engenders plausibility'.

Imperfect information and epistemic logic Finally, consider an *imperfect information* version of our game, where **E** does not know which move was played by **A**. This ignorance may arise for several reasons, as in real games — either private or public. Perhaps **E** did not pay attention to **A**'s move, or the latter was hidden from her (as with initial deals in a game of cards). Her uncertainty is graphically indicated by the dotted line in the following picture:

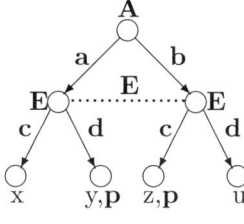

Figure 3

This picture suggests extending our language to an *epistemic logic* with knowledge operators, expressing things at **E**'s turn like

$K_\mathbf{E}(\langle a\rangle p \vee \langle b\rangle p)$ **E** knows she has a move guaranteeing **p**
$\neg K_\mathbf{E}\langle a\rangle p \ \& \ \neg K_\mathbf{E}\langle b\rangle p$ but she does not know *which one* does.

2.3 Intensional logic as usual

Logical languages of these various kinds express assertions about players' actions and their effects, which can be checked against game models in the usual semantical way. In addition to this model-checking mode, one can study proof systems axiomatizing valid principles for reasoning in special game classes, satisfying constraints such as unique-valuedness of moves, 'rationality' of behaviour under preferences, or 'perfect recall' in imperfect information games. This business-as-usual makes game logic into a rich form of applied intensional logic (cf. Stalnaker (1999)).

In this paper, however, we wish to develop a number of new themes that arise on the process view of games. For convenience, extensive games will mostly be taken to be *finite trees*.

2.4 Major issues in process theories

Existing logical process theories show a wide variety, depending on one's viewpoint on the relevant dynamic structure. Here are some key features (van Benthem (1996); Harel et al. (2000)).

a) At one extreme, processes are black boxes with external input / output only — at another, one cares for all decisions and actions that go into their internal states. One measures such grain levels in terms of 'simulations' between process graphs. If all decisions are relevant, some form of *bisimulation* is needed — with only external behaviour in focus, one uses less discriminating notions of so-called *trace equivalence*.

b) The other side of the coin are *internal languages* for describing a process: modal ones if all actions are relevant, much coarser ones if only traces matter.

c) Instead of working inside, one can look from the outside, putting together whole processes by *natural operations* such as choice or sequential composition. This takes *external languages* of operations, as in process algebra (Fokkink (2000)).

d) Finally, processes come in two flavours. Finite processes correspond to bounded tasks, described by terminating programs. But natural infinite processes should run forever, such as operating systems. The logical setting for the latter is some form of temporal logic, or extensions of dynamic logic like the 'μ-calculus'.

All these issues apply immediately to games. As good logicians, we start by assuming perfect information, lifting this restriction later.

3 Levels of representation

3.1 Games: from actions to outcomes

In setting one's sights, there are various natural levels of representation for a game, ranging from local *actions* to global *outcomes*. Do you play soccer for the brilliant moves, or just for the score-board? The difference may be high-lighted by means of a 'propositional game'. The propositional Distribution law $p\wedge(q\vee r) \leftrightarrow (p\wedge q)\vee(p\wedge r)$ suggests a comparison between two 2-person games with choices for players **A** and **E**:

Are the following two games the same?

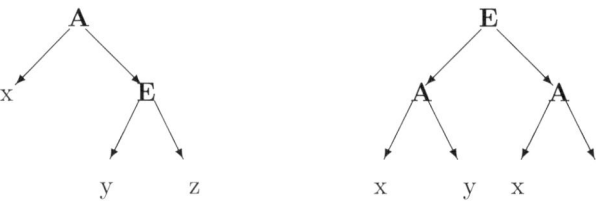

Figure 4

The intuitive answer is. "Yes, qua players' outcome control, No, qua intermediate actions." Clearly, the two games differ qua action structure. E.g., only on the left, but not on the right, **E** can be in a position to make the game end in both y and z. But if one analyzes available strategies, players' powers are indeed the same in both games depicted:

for **E**: {x, y}, {x, z} corresponding to her strategies "left" and "right"
for **A**: {x}, {y, z} for his strategies "left", "right" (on the left), and
⟨"left","left"⟩, ⟨"right","right"⟩ (on the right)

Note **A** also has powers {x, z} and {y, x} in the right-hand tree, corresponding to his strategies ⟨"left", "right"⟩ and ⟨"right", "left"⟩ — but these are just supersets of those given, and hence weaker powers.

The outcome level corresponds roughly to the earlier 'strategic form' of a game, but with the full enumeration of all strategies suppressed. (In the preceding example, the full strategy table is a 2 × 4 matrix.) The action level corresponds to the 'extensive form'. Later on, we shall also propose natural intermediate levels.

Now, as stated before, there are two styles of analysis for game representation. We can either define a *simulation* (a 'structural invariance'), or design a *description language* for the relevant game-internal properties. And these two styles can be matched precisely, as is well-known from logical process theories. The paradigmatic example is the match between 'bisimulation' of two processes, and their having the same properties definable in an internal *modal* language describing transitions. We can take this 'two-faced' style of analysis to games.

3.2 Actions in modal and dynamic logic

Expressive power of the language Extensive games were models for a modal or dynamic language which states, amongst others, 'strategic assertions'. In the game of Figure 1, player **E** had a strategy forcing outcome **p**:

$$[a \cup b]\langle c \cup d \rangle p$$

Thus, modal logic provides a systematic way of reasoning about players in games. A more general analysis of positions where **E** has such a strategy requires a modal *fixed-point formalism* like the μ-calculus. Let the assertion {**E**}p state that

> player **E** has a strategy for playing the game from
> now on which forces a set of outcomes all satisfying **p**.

This generic strategic assertion has the following definition:

$$\{\mathbf{E}\}p \leftrightarrow ((end\&p) \vee (turn_\mathbf{E} \& \vee_a \langle a \rangle \{\mathbf{E}\}p) \vee (turn_\mathbf{A} \& \&_a [a] \{\mathbf{E}\}p))$$

Reading this carefully should suffice for seeing the soundness of the equivalence. This definition can be used in model checking on finite game trees. In particular, unfolding the approximation stages of this fixed-point definition in the usual manner yields the well-known game-theoretic 'backward induction' algorithm determining the player with a winning strategy in finite zero-sum two-player games, or the 'subgame perfect Nash equilibrium' in general games of this sort.

These languages can also *define strategies* explicitly. E.g., **E**'s strategy in the above of

"play **right** if **A** played **left**, and play **left**, otherwise"

denotes a new relation on the game tree which can be defined as follows. Here we use standard dynamic logic operators of choice ∪ and composition ; together with a converse operator ∪ on actions, and a test operator ? on propositions:

$$((turn_\mathbf{E} \wedge \langle left^\cup \rangle turn_\mathbf{A})?;right) \cup ((turn_\mathbf{E} \wedge \langle right^\cup \rangle turn_\mathbf{A})?;left)$$

This allows us to reason about strategies much as about programs.

Bisimulation For modal-dynamic languages, the process equivalence is *bisimulation*, whose definition we take for granted here. Here are two key results on its match with the modal language:

Modal Invariance Lemma
For finite models, the following are equivalent:

(a) **M**, s is *bisimilar* to **N**, t

(b) **M**, s and **N**, t satisfy the same modal formulas.

State Definition Lemma
For each finite **M**, s, there exists a formula of propositional dynamic logic which holds only in the process graphs **N**, t bisimilar to **M**, s.

Bisimulation as it stands seems too fine an invariance for games, witness the following example of two non-bisimilar process graphs, viewed from their roots, with different arrows for different moves:

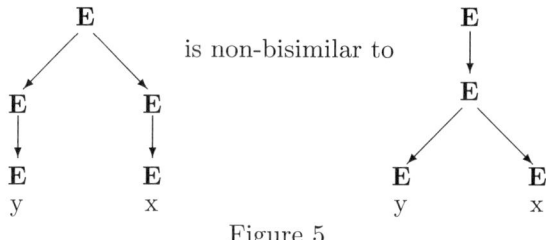
Figure 5

One would not normally distinguish these trees as different games — although even this structure sensitivity has occasionally been considered in game theory. However this may be, the above standard modal results are suggestive for what may be achieved for games as well. E.g., the second Lemma on complete bisimulation invariants (cf. Barwise and Moss (1997)) is reminiscent of Bonanno (1991)'s definition of characteristic 'game formulas' capturing the essence of a game.

3.3 Powers and a forcing language

Characterizing powers Now consider the 'powers' of players for forcing sets of outcomes, as in the example of the two games for propositional distribution. Here is how one describes this coarser level of game structure,

ρ_G^i sX *player i has a strategy for playing game G from state s whose resulting states are always in the set X*

Given this sense, it is reasonable to ask *closure under supersets* for these ρ's. Moreover, powers satisfy the following two 'logical' requirements:

Consistency   if **A** can force outcome set X and **E** set Y,
              then X, Y overlap
Determinacy   if **A** cannot force outcome set X,
              then **E** can force set −X

These three conditions capture this level of game analysis completely:

Fact Any two families of sets satisfying superset closure, consistency and determinacy are the complete lists of powers for the players in some two-person game.

Simulation and language The above suggests a *power bisimulation* between games G, H, played over two 'game boards' whose states need not be identical, only 'comparable'. A power bisimulation generalizes an ordinary modal bisimulation to a relation E between states in two models satisfying

1. *Local Harmony*: if x E y , then x, y satisfy the same atomic proposition letters.

2. *Outcome zigzag* for each player **i** :

    2a) if x E y and $\rho^i_G x$, U , then there exists a set V with $\rho^i_H y$, V and $\forall v \in V\ \exists u \in U\ $ u E v;

    2b) vice versa.

A matching modal language for this has a main operator

$\mathbf{G}, s \models \{\mathbf{i}\}\phi$ *player i has a strategy for playing game G starting from the current node s which guarantees an outcome state satisfying* ϕ.

This language can express the above three characteristic conditions on power sets as axioms. E.g., superset closure is the monotonicity axiom $\{\mathbf{i}\}\phi \to \{\mathbf{i}\}\phi \vee \psi$. With its power bisimulation, we then get the same results as those for modal and dynamic logic in Section 3.2 (cf. van Benthem (2001a)).

3.4 Intermediate game levels

Modal bisimulation and power bisimulation are two points on a spectrum of game equivalences. But the best options may lie in between. Sometimes we want to describe some of what happens at intermediate stages in a game. E.g., the dynamics of the two Distribution games was quite different. In the root on the left, but not on the right, player **A** can hand player **E** a choice between achieving q and r. This might be expressed in a variation on the preceding notation:

$$\{\!\{\mathbf{A}\}\!\}(\{\!\{\mathbf{E}\}\!\}q\ \&\ \{\!\{\mathbf{E}\}\!\}r),$$

true in the left root, false on the right. Here we read $\{\!\{\mathbf{E}\}\!\}\phi$ as

> player **E** has the power to take the game to some state, final *or intermediate*, where the proposition ϕ holds.

This is a natural intermediate level in the simulation/language spectrum, which supports definability results like before. It does not keep track of specific actions (as modal bisimulation does), but unlike power bisimulation, it does care about the dynamics of players 'on the move'.

Intermediate notions of game equivalence are investigated in van Benthem (2001b). Other candidates include 'alternating bisimulation', which treats maximal *control zones* of the game where the same player is to move via power bisimulation, while acting like a modal bisimulation when a *turn switch* occurs. This corresponds to using a dynamic language whose moves are sequences of actions inside a zone, ending in either the border of a new zone, or an endpoint.

All these simulations will also work on much more general graph representations for games than the finite trees assumed for our exposition. In particular, they also work on external 'game boards' suppressing game-internal predicates like 'turns' or 'wins'. Our more general conclusion is as follows. *Game equivalences are like process equivalences, and issues of appropriate language design are similar, too.* Now, there are also many notions of equivalence in the game-theoretic literature itself, often in the form of 'transformations' purporting to preserve relevant structure. Examples are inserting a redundant 'stalling move' into a game, or 'coalescing moves' made by the same player — which do exactly what their names say. These still remain to be compared with the bisimulation approach.

4 Game operations and game algebra

One can also work 'from the outside', and describe compositional operations that construct games out of games. This is another style of analyzing game structure, high-lighting algebraic properties.

4.1 Game operations

Natural operations on games can be extracted from various special sources (cf. Parikh (1985)). Examples are sequential operations like

choice	for either player between two games	$G \cup H$, $G \cap H$
role switch	for the two players in a game	G^d
composition	leading to successive play of games	$G;H$

with straightforward formal definitions as operations on game trees. Starting from one-move games, these build all finite game trees without simultaneous actions. But there are also parallel game compositions. E.g.

simultaneous play of two games $G \times H$

may be viewed as (a subset of) direct products in the model-theoretic sense: all pairs of states from G, H with moves being pairs of relevant moves in both components. Other parallel game operations are studied in linear logic (Abramsky (1995)), with interleaving of moves in different games and players switching between these. For a more general account of these game logics, cf. van Benthem (2000).

4.2 Game algebra

Game operations suggest game algebra. E.g., intuitively, choices for the two players are related by a De Morgan duality under role switch:

$$G \cap H = (G^d \cup H^d)^d$$

Or, typically, the left-distribution law for composition over choice

$$(G \cup H); K = (G;K) \cup (H;K)$$

holds for games, while right-distribution G;(H∪K) = (G;H)∪(G;K) does not — as the choice on the left-hand side (but not on the right-hand side) may depend on the outcome of first playing a game G. Another typical example is one more intuitively valid role switching principle:

$$(G;H)^d = G^d; H^d$$

These intuitions may be made more precise using the notions of Section 3.3. Given any game board with hard-wired powers for atomic games, one can easily compute power relations inductively for complex games constructed using choice, dual, and composition. Then two game expressions may be called *equivalent* if they have the same power relations for their players on all game boards. This validates the preceding observations, and many more. The complete algebra of the above sequential game operations has recently been axiomatized in Goranko (2000). The same line of thought extends to many other game operations. E.g., it is easy to see that the above direct product of games distributes over choice in both its arguments.

4.3 Languages

External game languages can be diverse. Extending the pure equational formalism, one can import complex game terms G into the earlier internal languages. E.g., a modality $\{G, \mathbf{i}\}\phi$ describes player $\mathbf{i}$'s powers in game G, relating them to powers in subgames. Here are some axioms of the resulting 'dynamic game logic' (Parikh (1985), Pauly (2001)):

$$\{G \cup H, \mathbf{i}\}\phi \leftrightarrow \{G,\mathbf{i}\}\phi \vee \{H,\mathbf{i}\}\phi$$
$$\{G;H, \mathbf{i}\}\phi \leftrightarrow \{G,\mathbf{i}\}\{H,\mathbf{i}\}\phi$$
$$\{G^d, \mathbf{i}\}\phi \leftrightarrow \{G,\mathbf{j}\}\phi \quad \text{(where } \mathbf{j} \text{ is the opposite player)}$$

5 From finite to infinite games

Infinite games also occur in game theory, e.g., with evolutionary behaviour. They also arise at various places in logic: e.g., evaluation games for fixed-point languages, or model comparison games for potential isomorphism can go on forever (cf. Hodges (1998)). Likewise, games can be infinite in computer science: e.g., with infinite repetitions of finite 'graph games', or with games modeling unlimited interaction. Both sorts may occur in the same setting. One can think of natural language as a series of finite terminating games, for episodes of interpreting a statement or engaging in argument, but all enabled by the never-ending game of Discourse which sets the rules of proper communication.

Intuitively, the outcomes of infinite games are not end states, but *infinite runs*. Accordingly, the atmosphere changes. Desired properties that players are aiming for include long-term features of these runs, such as 'safety' (no catastrophes have occurred) and 'fairness' (all participants were 'served' and treated correctly). Technically speaking, this can be formulated in extensions of the above modal formalisms, such as the mentioned μ-calculus. But it is often natural to switch to another formalism here, viz. branching *temporal logic*. We will not develop this alternative viewpoint here, but we do present one example that shows the line of thinking. One good design methodology for reasonable game languages is the following: *look at the minimum needed to formalize natural game-theoretic proofs.*

Consider the well-known Gale-Stewart Theorem, which says that infinite games with an *open* winning set for one of their players are *determined*: i.e., one player has a winning strategy. Its proof hinges on a key observation which one can easily verify intuitively for all games:

> Fact   Either player **E** has a winning strategy, or player **A** has a strategy forcing only branches on which player **E** never gets a winning strategy.

The straightforward formalization of the lemma is in a *temporal logic*, evaluating formulas in branching time models **M** at ordered pairs ⟨h,t⟩ of the current branch h , and point t on it. Then we have the following:

$$\{G,\mathbf{E}\}\phi \vee \{G,\mathbf{A}\}\ A\ \neg\{G,\mathbf{E}\}\phi$$

with $\{G,\mathbf{i}\}\phi$ again the game modality stating that there exists a strategy for player **i** ensuring that only runs result having the current history h up to point t as an initial segment, which satisfy the temporal logic formula ϕ. The temporal logic comes in explicitly through the operator A in the right-hand disjunct. This is the standard operator expressing "always on the current branch". Incidentally, this valid principle of temporal game logic is a weak universal form of determinacy — while real determinacy only arises for special winning conventions ϕ.

Some people are fond of construing ideological conflicts between 'temporal logic' and 'dynamic logic' approaches to processes. But in the light of eternity (the best light there is), *both* have their uses and the same is no doubt true with games.

6 Coping with imperfect information

Here is the first real test case for the preceding analysis when confronted with more realistic game theory. What happens to pure logics of perfect information when they meet the reality of *imperfect information* games, such as card games, or the under-informed circumstances in which people must act?

6.1 From perfect to imperfect information

As an example, consider a rare case of imperfect information inside logic. Hintikka and Sandu (1997) let a formula

$$\forall x\ \exists y/\mathbf{x}\ \neg x=y$$

describe an evaluation game for the formula $\forall x\ \exists y\ \neg x=y$ where the Verifier player **E** does not know which challenge x was made by the Falsifier player **A**, and must respond with a witness y different from x all the same. Real game examples are in Osborne and Rubinstein (1994), and van Ditmarsch (2000). Game trees are now annotated with dotted lines indicating players' uncertainties. Recall the game tree of Figure 3. In this tree, player **E** does not know the initial move played by **A**. For instance, think of a two-object 'game board' for the formula game $\forall x\ \exists y/\mathbf{x}\ \neg x=y$, where y, z are the 'winning positions' for **E**. Unlike finite perfect information games, these games can be *non-determined*, witness our example. **A** has no strategy forcing the game to end in the set {x, u}, but **E** has no *uniform strategy* forcing the complement set {y, z} either. A uniform strategy is a strategy prescribing the same action for a player at all states that she cannot distinguish — as opposed to strategies whose instructions depend on things players do not know in this setting.

These games may look more mysterious than the perfect information games of logic and computer science. A good test question is this: *when are two imperfect information games the same?* Many people claim, e.g., that the above game ∀x ∃y/**x** Rxy is 'obviously equivalent' to ∃y ∀x Rxy. But the 'Thompson transformations' presented in Osborne and Rubinstein (1994) make it 'equivalent to'

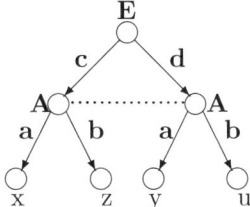

Figure 6

Thus, ∀x ∃y/**x** Rxy would turn into ∃y ∀x/**y** Rxy. Who is right?

6.2 Uniform outcome equivalence and a forcing modality

Reversing the order of Section 3, let us first consider players' powers in games like this. Call imperfect information games equivalent when players can force the same sets of outcomes via *uniform strategies*.

Example   In Figures 3 and 6, players have the same uniform powers:

   E can force {x,z}, {y,u}   **A** can force {x, y}, {z, u}

Indeed, game equivalence in terms of powers via uniform strategies sanctions all 'Thompson transformations'. For concreteness, we also add a normal form result (van Benthem (2000)) which demonstrates the 'collapsing' effects of this notion.

Observation   Any two families of sets satisfying Monotonicity and Consistency are realized in a *two-move* imperfect information game.

Example   Powers for player **A** : {A,B,C}, {C,D} , **E** : {B,C}, {A,D}. These are realized in the game of Figure 7.

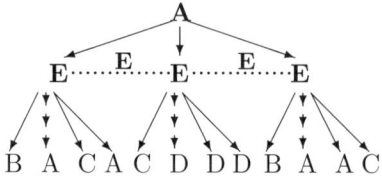

Figure 7

As in Section 3, the uniform power analysis suggests a matching internal language for describing games with a strategy modality

$\{G, i\}\phi$ stating that *player i has a uniform strategy forcing a set of outcomes all of which satisfy proposition ϕ*

Expressiveness and validity for this language can be studied just as before. Moreover, staying with logical matters, these game equivalences also suggest an extended algebraic calculus, with finite trees plus dotted lines as formulas of some sort. This is a generalized version of the Game Algebra of Section 4, allowing for independencies between stacked operators. No explicit axiomatization of this system appears to be known. Likewise, in computer science, no imperfect information versions are known of Process Algebra or Linear Logic.

6.3 Actions and information: dynamic-epistemic logic

Much natural game structure is not preserved in the above way. Recall the plight of player **E** described in Section 2.2. In the game of Figure 3, we want to say intuitively that after move b has been played, **E** *knows* that playing either c or d will give her a win in the set {x, y}. That is, in the middle nodes of the game tree, the following combined dynamic-epistemic formula holds:

$$K_{\mathbf{E}}(\langle a \rangle \text{win} \vee \langle b \rangle \text{win})$$

Thus, **E** knows 'de dicto', in the philosopher's sense, that she has some winning move. But 'de re', there is no specific move of which she knows that it is winning:

$$\neg K_{\mathbf{E}} \langle a \rangle \text{win} \ \& \ \neg K_{\mathbf{E}} \langle b \rangle \text{win}$$

Thus game trees become models for a *dynamic-epistemic* language with arbitrary actions and S5 uncertainty relations. Upon addition of a *common knowledge* operator, it can analyze many standard features of imperfect information games: including common knowledge of 'who is to move', or 'which moves are available'. In this format, uniform strategies are not just winning, but to a first approximation the ones whose players know that they are winning. The general logic of these models is easily axiomatized, since it has no significant interaction principles between action and knowledge. But special styles of playing do. Van Benthem (2001a) has dynamic-epistemic analyses of 'Perfect Recall', 'Bounded Memory', plus uniform strategies, and mechanisms for information update in imperfect information games.

7 Preferences and rational behavior

Even more briefly, let us consider the other main challenge of game theory. Games model much more interesting behaviour than two-valued attitudes 'win' versus 'lose' when we encode players' more fine-grained *preferences*, say via their values for outcomes. Here is just a bit of the flavor of what happens then.

7.1 Game values, equilibrium and backward induction

Figure 2 showed our perennial game tree indicating how players evaluate outcomes. As we saw in Section 2.2, the predicted equilibrium outcome in this case is y, where both players get value 1. But players need not always get their best value along these lines. As another illustration, consider the two games for the distribution law in Figure 4. Assume the following preference values:

A	x:1	y:0	z:2
E	x:0	y:2	z:1

By some straightforward reasoning about utility maximization in the game tree, the predicted equilibrium outcome on the left is x (value (1,0)), while on the right, it is z (value (2,1)). The well-known *backward induction algorithm* of game theory — also known under various names in computer science — computes such equilibrium outcomes systematically. Starting from the leaves, it assigns values for each player to each node. Here is the rule:

> Suppose **E** is to move, and all values for daughters are known. The **E**-value is the maximum of all the **E**-values on the daughters, the **A**-value the minimum of the **A**-values at all **E**-best daughters. The dual case for **A**'s turns is completely analogous.

We display the computed pairs (**A**-value, **E**-value) in the last example:

The top values match the predicted outcomes: x on the left, z on the right. One description of what the backward induction algorithm produces for any game G is a subtree $\mathcal{E}(G)$, giving just the *best play*. This can be encoded as a subset of the original nodes, or as a collection of 'best moves'. The algorithm actually gives more information than just this tree from the root. En route, it also computes best moves from positions that will not be actually reached in $\mathcal{E}(G)$.

Remark This algorithm is not neutral. It embodies a defeasible assumption about players' rational behaviour: that they choose randomly among their best options. Different assumptions yield different algorithms. With a random opponent, one must minimize over all possible outcomes of her actions. With a benevolent opponent one may reckon with the maximal outcome from among her best moves. Thus, the mathematics of these game trees depends on rationality assumptions.

7.2 Logic issues revisited

Our previous considerations still make sense in this setting, but formulations may branch into several options. Here is just a sketch (cf. van Benthem (2001b) for details). First consider our test question of appropriate *game equivalences*. Here are some possible answers:

(a) Factor out rationality assumptions. Let $\mathcal{E}(G)$ be whatever our preferred solution algorithm says, and just apply the earlier equivalences of Section 3 to these 'best games'!

(b) Demand equality of values as computed by our favourite algorithm: that is, the equilibrium outcomes are the same.

(c) Consider powers as computed in Section 3, take their value for players to be the minimum, and demand that the maximum over these is the same.

Each represents a possible way of thinking about game equivalence. Next, consider *logical languages* for this richer setting. Again, there are several options, from simpler to more complex. Mathematically, the backward induction axiom computes a recursive binary relation of 'relative plausibility':

$u >^E v$ iff *E's expected value at u is greater than that at v*

This relation has an obvious inductive definition, which can be stated in a modal fixed-point language. But for many purposes, the only thing that matters is just the 'best next move' relation, which supports unary modalities of *belief* about further development of the game. When combined with imperfect information and epistemic modalities, such languages provide a rich medium for representing players' reasoning in the course of real games, including counterfactual statements and *belief revision* (cf. Stalnaker (1999), Board (2001)).

Finally, as for *game algebra*, things get much more complicated. E.g., the earlier distribution law for ∪, ∩ now fails, as its two sides got different best outcomes. Game algebra with preferences is therefore another intriguing open area.

8 Conclusion

We have shown how games may be analyzed in the style of logical process theories, starting from the pure action case with perfect information, and then — more speculatively — including realistic features like imperfect information and preferences. This analysis does not solve any of the outstanding problems of game theory, or logic, but it does raise some interesting new questions. And it influences work both ways. E.g., one can introduce imperfect information or preferences into standard logic games, and get a whole range of new issues. But also, issues of game algebra and infinite games seem a natural addition to the standard repertoire of game theory. So, to return to our opening examples: what does this paper have to say about players meeting in Stag Hunts and Centipedes? Nothing per se. But we did have something to say about another type of meeting. There is no denying that the research agendas of logicians and game theorists are not the same. But they do overlap enough to make their encounters meaningful.

References

Abramsky, S. 1995. *Semantics of Interaction*. Lecture notes, Department of Computer Science, University of Edinburgh.

Barwise, J. and L. Moss. 1997. *Vicious Circles*. CSLI Publications, Stanford.

van Benthem, J. 1996. *Exploring Logical Dynamics*. CSLI Publications, Stanford.

van Benthem, J. 2000. *Logic in Games*. Electronic lecture notes, http://staff.science.uva.nl/~johan/, and manuscript versions 2001, 2002, ILLC Amsterdam.

van Benthem, J. 2001a. Dynamic-Epistemic Logic of Games. In Bonanno, G. and W. van der Hoek, editors, *Bulletin of Economic Research*, volume 53, pages 219–248.

van Benthem, J. 2001b. Extensive Games as Process Models. In Pauly, M. and P. Dekker, editors., special issue of *Journal of Logic, Language and Information*, volume 11, pages 289–313.

Blackburn, P., M. de Rijke, and Y. Venema. 2001. *Modal Logic*. Cambridge University Press, Cambridge.

Board, O. 2001. Dynamic Reasoning about Knowledge and Belief. *Department of economics, Oxford University, talk at IMGTA, Ischia 2001*.

Bonanno, G. 1991. The Logic of Rational Play in Games of Perfect Information. *Economics and Philosophy 7, 37-65*.

van Ditmarsch, H. 2000. *Knowledge Games*. Ph.D. thesis, ILLC Amsterdam and Department of Technical Cognitive Sciences, Groningen.

Fokkink, W. 2000. *Introduction to Process Algebra*. Texts in Theoretical Computer Science, Springer-Verlag, Berlin.

Goranko, V. 2000. *The Basic Algebra of Game Equivalences*. Report PP-2000-12, Institute of Logic, Language, and Computation, Amsterdam.

Harel, D., D. Kozen, and J. Tiuryn. 2000. *Dynamic Logic*. MIT Press, Cambridge (Mass.).

Hintikka, J. and G. Sandu. 1997. Game-Theoretical Semantics. In van Benthem, J. and A. ter Meulen, editors, *Handbook of Logic and Language*, pages 361–410. Elsevier, Amsterdam.

Hodges, W. 1998. *An Invitation to Logical Games*. Lecture notes, Department of Mathematics, Queen Mary's College, London.

Osborne, M. and A. Rubinstein. 1994. *A Course in Game Theory*. MIT Press, Cambridge (Mass.).

Parikh, R. 1985. The Logic of Games and its Applications. *Annals of Discrete Mathematics 24*.

Pauly, M. 2001. *Logic for Social Software*. Ph.D. thesis, ILLC and CWI Amsterdam.

Stalnaker, R. 1999. Extensive and Strategic Form: Games and Models for Games. *Research in Economics, 53:293-291*.

Informationally Independent Connectives

Gabriel Sandu and Ahti Pietarinen

> In this paper we consider some conceptual problems raised by Wilfrid Hodges in connection with Independence-friendly languages, that is, languages in which sentences receive a semantic interpretation in terms of games of imperfect information. Hodges pointed out that there are certain anomalies which arise when sentential connectives are interpreted as informationally independent. We shall investigate the nature of this phenomenon in a more general setting, by looking more closely at the relation between semantical games of imperfect information and games in classical game theory (von Neumann and Morgenstern).

1 Sentential logic and games

The overall purpose of this paper is to make more precise the insight that sentential logic can be given a game-theoretic content. That is, in a sentence like

$$(A \wedge B) \vee (C \wedge D)$$

we may think of disjunction as prompting a choice between the two disjuncts (**Left** or **Right**), and similarly for conjunction. Naturally, choices depend on and are made possible by earlier choices. Thus in our particular example, the effective choice between C and D becomes possible only if there were an earlier choice of the right disjunct. So far so good: all this has been known since the early seventies.

The idea of associating games with sentential logic suggests some further generalisations. If choices *depend* on other choices, it is also natural to think of some choices being *independent* of others. For instance, in the above example we may think of a player choosing **Right** in the first move, and then of a second player having to choose between **Left** and **Right** in the second move, except that now the second player

---

*Address (both authors): Department of Philosophy, University of Helsinki, P.O. Box 9, FIN-00014 University of Helsinki, Finland.

does not 'know' the earlier choice. This way of looking at connectives gives rise to a nice extension of propositional logic which has been studied elsewhere (see Sandu and Pietarinen 2001). The phenomenon of informational independence for quantifiers has been introduced by Henkin (1961). Its extension, however, to propositional connectives is full of death traps. Let us start with one of them.

As to the particular example which is our concern here, one may ask: how is the second player supposed to know that it is his turn to move, without knowing the previous choice? For instance, if the second player is supposed to know that he has to choose between C and D, then he can infer that the first player has chosen the right disjunct. And if he is supposed to choose between A and B, then he can infer that the previous choice was **Left**. As we see, the idea of informational independence brings in some interesting complications which involve the notion of players' information in the game, their knowledge of the game, and so on. Some of these complications have been pointed out by Hodges (1997). We will try to sort them out in what follows. In order to do that we shall look at similar issues in the theory of games, in particular in the theory of games with imperfect information. After that, we shall formulate a game as a Kripke structure, and in the end we shall transfer all that setting to games involving informationally independent connectives.

1.1 Extensive games of perfect information

We fix a set of actions A which represents the set of possible choices of the players in the game. A sequence $(a_1 \ldots a_n)$ of actions represents the consecutive choices of the players, $a_i \in A$.

Definition 1.1 *An extensive game $\mathcal{G}_A$ of perfect information is a tuple*

$$\mathcal{G}_A = \langle N, H, Z, P, (u_i)_{i \in N} \rangle$$

such that

1. *N is the set of players of the game;*

2. *H is a set of sequences of actions from A, which are called histories, or plays of the game. We require that:*

    (a) the empty sequence $()$ is in H;

    (b) If $h \in H$, then any initial segment of h is in H too;

3. *Z is the set of maximal histories of the game;*

4. $P\colon H \setminus Z \to N$ *is the player function which assigns to every non-terminal history the player whose turn is to move;*

5. *each u_i is the payoff function for player $i \in N$, that is, a function which specifies for each maximal history what is the payoff for player i.*

For any non-terminal history $h \in H$ we define
$$A(h) = \{x \in A \mid h \frown x \in H\}$$
A strategy for a player i is any function
$$f_i\colon P^{-1}(\{i\}) \to A$$
such that $f_i(h) \in A(h)$, where $P^{-1}(\{i\})$ is the set of all histories where player i is to move.

From the class of extensive games of perfect information, we single out a particular subclass, which is the class of *zero-sum* (*win-loss*) games. These are games played by two players, that is, $N = \{\exists\ (Eloise), \forall\ (Abelard)\}$, and in addition:

- $u_\exists(h) = -u_\forall(h)$ (the game is strictly competitive), for all terminal histories h.

- $u_\exists(h) = 1$ or $u_\exists(h) = -1$ (that is, $\exists$ either wins or loses), for all terminal histories $h \in H$.

The following theorem is an old result due to Zermelo:

Theorem 1.1 *Every finite extensive zero-sum game is determined: either player $\exists$ or player $\forall$ has a winning strategy in the game.*

Example 1.1 *The password game of perfect information:* Abelard *tells* Eloise *a password* R *or* L. *If* Eloise *is able to repeat it later on, she wins, and* Abelard *loses. Otherwise* Abelard *wins and* Eloise *loses. This is depicted in Figure 1.*

Eloise*'s winning strategy is very simple:* $f_\exists(\mathsf{L}) = \mathsf{L}, f_\exists(\mathsf{R}) = \mathsf{R}$.

1.2 Extensive games of imperfect information

The games are exactly as before, except that the players might not know what happened earlier in the game. A different way to say the same thing is that players may not distinguish between histories of the game. Consider the example of the password game above, except that

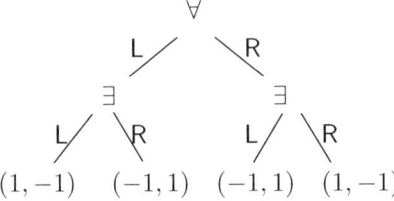

Figure 1: The password game of perfect information.

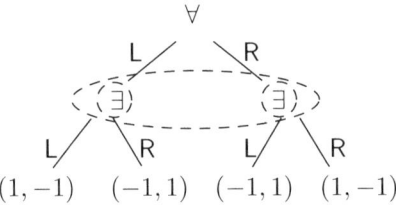

Figure 2: The password game of imperfect information.

now player ∃ does not know (or forgets) the password given to him by player ∀. That will mean that ∃ does not distinguish between the histories L and R, or alternatively that the two histories are equivalent for ∃, L $\sim_\exists$ R, a fact marked in Figure 2 by an oval around some nodes.

Actually, every single history is indistinguishable from itself, so in Figure 2 we also have indicated (by circles) the equivalence of the relevant histories to themselves. We then require that, for each player $p \in N$, the strategy function f_p be uniform on equivalent histories, that is, for $h, h' \in H$:

$$h \sim_p h' \implies f_p(h) = f_p(h').$$

The information about the equivalent histories in the game is primitive information which comes with the specification of the game.

Definition 1.2 *An extensive game $\mathcal{G}_A$ of imperfect information is a tuple*

$$\mathcal{G}_A = \langle N, H, Z, P, (\mathcal{I}_i)_{i \in N}, (u_i)_{i \in N} \rangle$$

such that the sets $N, H, Z, P,$ and $(u_i)_{i \in N}$ are exactly as before, and for each player $i \in N$, the set $\mathcal{I}_i$ is a partition of the set $\{h \in H \mid P(h) = i\}$. Each member $I_i \in \mathcal{I}_i$ is called an information set for player i.

Thus the new element here is the collection $(\mathcal{I}_i)_{i \in N}$ of information partitions.

The requirement of the uniformity of the strategy functions of the players can now be expressed by

$$h, h' \in I_i \in \mathcal{I}_i \Longrightarrow f_i(h) = f_i(h'), \text{ for } i \in N.$$

In the password example above, we have the following partition:

$$\mathcal{I}_\exists = \{\{\mathsf{L},\mathsf{R}\}\}; \quad \mathcal{I}_\forall = \{\{()\}\};$$

Imperfect information does at least two things:

- It introduces indeterminacy in the game.
- It allows for a phenomenon known in game theory as *signalling*.

The password game in the picture above provides an example of indeterminacy: neither player $\exists$ nor player $\forall$ has a winning strategy in the game.

An example of signalling is provided by a variant of the password game.

Example 1.2 *The extended password game. There are two teams: the team $\forall$ (consisting of one player,* Abelard*) is playing off against the team $\exists = \{\exists_1, \exists_2\}$. The game is played in the following way: Abelard tells player $\exists_1$ a password* L *or* R *(without player $\exists_2$ hearing it), after which player $\exists_1$ tells a password* L *or* R *to player $\exists_2$. Finally, if player $\exists_2$ is able to repeat the password told by* Abelard*, the team $\exists$ wins; otherwise the team $\forall$ wins. This game is drawn in Figure 3.*

We treat each team as one player. The partitions of the game are:

$$\mathcal{I}_\exists = \{\{\mathsf{L}\}, \{\mathsf{R}\}, \{(\mathsf{L},\mathsf{L}), (\mathsf{R},\mathsf{L})\}, \{(\mathsf{R},\mathsf{R}), (\mathsf{L},\mathsf{R})\}\}; \quad \mathcal{I}_\forall = \{\{()\}\}.$$

Notice what is going on here: Although player $\exists_2$ does not 'see' the choice made by player $\forall$, player $\exists_1$ can reveal it to her. This provides the team $\exists$ with a winning strategy, consisting of two functions, one for each member of the team:

$$f_{\exists_1}: \{\mathsf{L},\mathsf{R}\} \to \{\mathsf{L},\mathsf{R}\}, f_{\exists_2}: \{\mathsf{L},\mathsf{R}\} \times \{\mathsf{L},\mathsf{R}\} \to \{\mathsf{L},\mathsf{R}\}$$

defined by

$$f_{\exists_1}(\mathsf{L}) = \mathsf{L}, f_{\exists_1}(\mathsf{R}) = \mathsf{R}$$
$$f_{\exists_2}(\mathsf{L},\mathsf{L}) = f_{\exists_2}(\mathsf{R},\mathsf{L}) = \mathsf{L}, f_{\exists_2}(\mathsf{L},\mathsf{R}) = f_{\exists_2}(\mathsf{R},\mathsf{R}) = \mathsf{R}.$$

Notice that, for any $\mathsf{S} \in \{\mathsf{L},\mathsf{R}\}$*:*

$$u_{\exists_1}(\mathsf{S}, f_{\exists_1}(\mathsf{S}), f_{\exists_2}(\mathsf{S}, f_{\exists_1}(\mathsf{S}))) = u_{\exists_2}(\mathsf{S}, f_{\exists_1}(\mathsf{S}), f_{\exists_2}(\mathsf{S}, f_{\exists_1}(\mathsf{S}))) = 1.$$

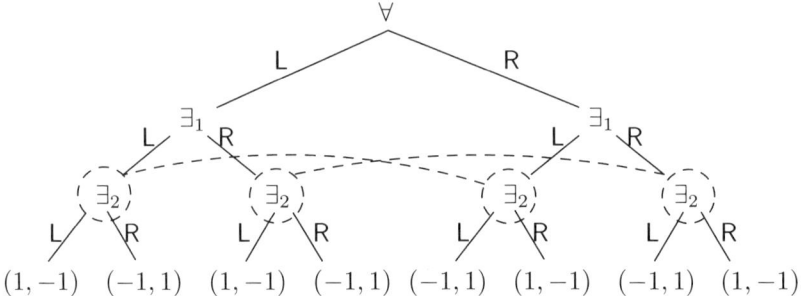

Figure 3: The extended password game.

Remark 1.1 *We can also take an extensive game $\mathcal{G}_A$ of perfect information to have the general form $\mathcal{G}_A = \langle N, H, Z, P, (\mathcal{I}_i)_{i \in N}, (u_i)_{i \in N} \rangle$, where each $\mathcal{I}_i$ is a partition of the set $\{h \in H \mid P(h) = i\}$ defined by:*

$$\mathcal{I}_i = \{\{h\} \mid h \in H, P(h) = i\}.$$

Very little is known, to the best of our knowledge, about games of imperfect information. Looking at the literature, one encounters several constraints which may be put on the partition sets $\mathcal{I}_i$. Here are a few of them.

The consistency condition

This can be stated as follows.

$$\text{For all } h, h' \in H : h, h' \in I_i \in \mathcal{I}_i \implies A(h) = A(h').$$

(cf. Osborne and Rubinstein 1994, p.200).

The idea behind this condition is quite simple: if a player does not distinguish between two histories h and h', then the choices available to him after h must be the same as those available to him after h'. If that was not so, that is, if $A(h) \neq A(h')$, then the player could distinguish between h and h', contrary to the assumption. In other words, *to indistinguishable histories there should exist corresponding indistinguishable futures*. All the games analysed so far satisfy this condition.

The conditions in the next group put constraints on the past of indistinguishable histories: *to indistinguishable histories there should exist corresponding indistinguishable pasts*.

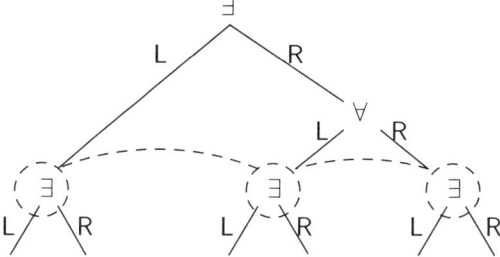

Figure 4: A game of imperfect information

The von Neumann & Morgenstern condition

This condition states that

$$h, h' \in I_i \in \mathcal{I}_i \implies length(h) = length(h'), \text{ for all } i \in N.$$

(cf. Bonanno and Battigalli 1997, p.241).

This condition is automatically satisfied for all games of perfect information, but it excludes game frames of imperfect information like the one in Figure 4.

In this example, L, (R, L), and (R, R) all belong to the same information set, but $length(\mathsf{L}) \neq length(\mathsf{R},\mathsf{L})$. The *von Neumann & Morgenstern condition* can also be seen to exclude the *absent-minded driver* game, not excluded by the *consistency condition*. In this game, the driver, in order to go home, has to take the highway and get off at the second exit. Turning at the first exit (E_1) gets him the payoff 0, and turning at the second exit (E_2) gets him the maximum payoff 4. If he continues (C_2) after the second exit he will get into a long road from which it will take him quite a while to get back home (payoff 1). The driver is absent-minded in the sense that when reaching an intersection, he does not remember how many intersections he has passed (see Rubinstein 1998). The situation is described in Figure 5.

The condition of perfect recall

This condition requires that the players remember their earlier moves. Fix a game $\mathcal{G}_A = \langle N, H, Z, P, (\mathcal{I}_i)_{i \in N}, (u_i)_{i \in N} \rangle$ of imperfect information. We define, for every history $h \in H$, the experience $\exp_i(h)$ of player i along the history h as the sequence consisting of the information sets that the player encounters in the history h together with the actions he takes at them.

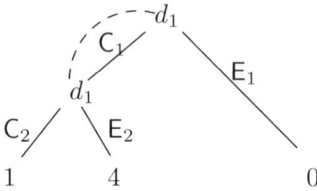

Figure 5: The absent-minded driver game.

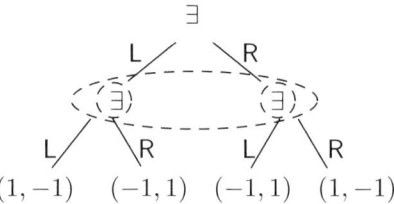

Figure 6: A game of imperfect information

The game $\mathcal{G}_A = \langle N, H, Z, P, (\mathcal{I}_i)_{i \in N}, (u_i)_{i \in N} \rangle$ is of perfect recall if for every player $i \in N$, and $h, h' \in H$:

$$h, h' \in I_i \in \mathcal{I}_i \implies \exp_i(h) = \exp_i(h').$$

This condition rules out games of imperfect information such as the one drawn in Figure 6. In this game, $\exp_\exists(\mathsf{L}) = \{(), \mathsf{L}\} \neq \exp_\exists(\mathsf{R}) = \{(), \mathsf{R}\}$, but $\mathsf{L}, \mathsf{R} \in \{\mathsf{L}, \mathsf{R}\} \in \mathcal{I}_\exists$.

1.3 Extensive forms as Kripke frames

Let $\mathcal{G}_A = \langle N, H, Z, P, (\mathcal{I}_i)_{i \in N}, (u_i)_{i \in N} \rangle$ be an extensive game (of perfect or imperfect information). We can extract from it a *Kripke frame* $F_{\mathcal{G}_A} = \langle W, (R_i)_{i \in N} \rangle$, where W is a set of possible worlds, and each R_i is an alternativeness relation for player i. This is done in a straightforward way:

- $W = H$,
- $R_i = \mathcal{I}_i \cup \{\{h\} \mid h \in H, P(h) \neq i\}$.

Clearly, every set R_i is a partition of the whole set H. Accordingly, every history $h \in H$ belongs to a member R_i^h of R_i. The set R_i^h may be thought of as the set of alternatives of h for player i. Alternatively, we may think of each R_i^h as the information i has at the possible world h.

The following two properties are trivially satisfied:

P1 $h \in R_i^h$, for every $h \in W$.

P2 $h' \in R_i^h \to R_i^{h'} = R_i^h$.

(cf. Osborne and Rubinstein 1994, p.69.)
For any $h, h' \in W$, let us define

$$R_i h h' \iff \text{there is } I_i \in \mathcal{I}_i, \text{ such that } h, h' \in I_i.$$

It is easy to see that the following are satisfied:

- $R_i h h$, for every $h \in W$.
- $R_i h h' \land R_i h' h'' \to R_i h h''$.
- $R_i h h' \land R_i h h'' \to R_i h' h''$.

From a Kripke frame F we obtain a *Kripke structure* S_F (based on the frame F) by adding a valuation function V, that is,

$$S_F = \langle W, (R_i)_{i \in N}, V \rangle,$$

where V associates with each primitive propositional atom p a set of possible worlds, $V(p) \subseteq W$. Intuitively, we may think of p as describing an event in the frame F. We then have the usual definition of the relation '$(S_F, h) \models A$' (the sentence A is true in the structure S_F at the possible world h):

1. $(S_F, h) \models p \iff h \in V(p)$
2. $(S_F, h) \models \neg A \iff \text{not } (S_F, h) \models A$
3. $(S_F, h) \models A \lor B \iff (S_F, h) \models A \text{ or } (S_F, h) \models B$
4. $(S_F, h) \models K_i A \iff \text{for all } h' \text{ such that } R_i h h': (S_F, h') \models A$.

In the password game of imperfect information, let p represent the event 'L has been chosen' and q the event 'R has been chosen'. That is, we let $V(p) = \{\mathsf{L}, (\mathsf{L}, \mathsf{L}), (\mathsf{R}, \mathsf{L})\}$, and $V(q) = \{\mathsf{R}, (\mathsf{L}, \mathsf{R}), (\mathsf{R}, \mathsf{R})\}$. Then in the corresponding Kripke structure $S_F = \langle W, (R_i)_{i \in N}, V \rangle$, with

$$W = \{(), \mathsf{L}, \mathsf{R}, (\mathsf{L}, \mathsf{L}), (\mathsf{L}, \mathsf{R}), (\mathsf{R}, \mathsf{L}), (\mathsf{R}, \mathsf{R})\};$$

$$R_\exists = \{\{()\}, \{\mathsf{L}, \mathsf{R}\}, \{(\mathsf{L}, \mathsf{L})\}, \{(\mathsf{L}, \mathsf{R})\}, \{(\mathsf{R}, \mathsf{L})\}, \{(\mathsf{R}, \mathsf{R})\}\};$$
$$R_\forall = \{\{()\}, \{\mathsf{L}\}, \{\mathsf{R}\}, \{(\mathsf{L}, \mathsf{L})\}, \{(\mathsf{L}, \mathsf{R})\}, \{(\mathsf{R}, \mathsf{L})\}, \{(\mathsf{R}, \mathsf{R})\}\};$$

and V as above, we have:

$$(S_F, \mathsf{L}) \models K_2 p, \ (S_F, (\mathsf{L}, \mathsf{L})) \models K_2 p, \ (S_F, (\mathsf{R}, \mathsf{L})) \models K_2 p.$$

Analogously for q. On the other side,

$$(S_F, \mathsf{L}) \models K_1(p \vee q), \ (S_F, \mathsf{L}) \not\models K_1 p, \ (S_F, \mathsf{L}) \not\models K_1 q.$$

The explicit representation of knowledge in a game allows us to formulate some extra conditions on the relation between the information of the players, and their knowledge of the game. Here are some of them:

1. The coherence between knowledge and information:

$$\mathcal{I}_i \subseteq R_i.$$

Notice that the way we have defined the sets R_i guarantees that this condition is automatically satisfied.

2. Players have knowledge about the conditions of the game. This condition consists of two sub-conditions:

2a. Players have knowledge about the payoffs of the game. Let τ_h represent the payoff of the terminal history $h \in W$, that is, the interpretation of τ_h is the set of maximal histories h which have the same payoff. Then our condition can be formulated as:

$$(S_F, s) \models K_i \tau_h, \text{ for every } i \in N \text{ and maximal history } s \in W.$$

2b. Players have knowledge about their choices. Let ζ_a represent the event of the action $a \in A$ being a choice at some history in the game, that is, the interpretation of ζ_a will be the set $\{h \in W \mid h \frown a \in W\}$. The present requirement becomes

$$(S_F, s) \models K_i \zeta_a, \text{ for every } i \in N \text{ and } s \in W \text{ such that } s \frown a \in W.$$

Notice that both requirements are automatically satisfied in the present setting because of the way in which the accessibility relations R_i have been defined. Actually, the last condition follows from the consistency.

2 Informationally independent connectives

We are now ready to go back to our original problem, and analyse some of the cases of informationally independent connectives.

2.1 Games of perfect information

Let φ be an ordinary propositional formula. A semantical game $\mathcal{G}(\varphi, M)$ played with the formula φ in the model M is played by two players, *Abelard* ($\forall$) and *Eloise* ($\exists$), and is defined in the following way:

- φ is an atomic formula: No move takes place. If $M \models \varphi$, then $\exists$ wins right away; otherwise $\forall$ wins.
- φ is $(\chi \vee \psi)$: player $\exists$ chooses a sentence $\theta \in \{\chi, \psi\}$, and the game goes on as in $\mathcal{G}(\theta, M)$.
- φ is $(\chi \wedge \psi)$: identical with the previous case, except that player $\forall$ chooses a sentence.
- φ is $\neg \psi$: the same as the game $\mathcal{G}(\psi, M)$, except that the players $\forall$ and $\exists$ are transposed, including the rules of winning and losing above.

A strategy for a player in the game is a function which yields a choice for each move of the respective player in the game.

We stipulate that:

- $M \models^+_{\text{GTS}} \varphi$, if and only if player $\exists$ has a winning strategy in $\mathcal{G}(\varphi, M)$;
- $M \models^-_{\text{GTS}} \varphi$, if and only if player $\forall$ has a winning strategy in $\mathcal{G}(\varphi, M)$.

The next results are straightforward.

Proposition 2.1

1. Player $\exists$ has a winning strategy in $\mathcal{G}(\varphi, M)$ if and only if player $\forall$ has a winning strategy in $\mathcal{G}(\neg \varphi, M)$;
2. Player $\exists$ has a winning strategy in $\mathcal{G}(\neg \varphi, M)$ if and only if player $\forall$ has a winning strategy in $\mathcal{G}(\varphi, M)$.

Proposition 2.2 *For every sentence φ, the game-theoretical truth (falsity) and the Tarski-type truth (falsity) coincide, that is:*
$$M \models^+_{GTS} \varphi \iff M \models_{Tarski} \varphi$$
$$M \models^-_{GTS} \varphi \iff M \not\models_{Tarski} \varphi.$$

We can present a semantical game $\mathcal{G}(\varphi, M)$ of perfect information as a zero-sum extensive game $\mathcal{G}_A = \langle N, H, Z, P, (u_i)_{i \in N} \rangle$ defined in the preceding section, where
$$A = \{\psi \mid \psi \text{ is a subformula of } \varphi\}.$$

H is formed inductively:

1. $() \in H$.

2. If $\varphi = \psi \vee \theta$ $(\psi \wedge \theta)$, then $\psi, \theta \in H$. In addition, $P(h) = \exists \ (\forall)$.

3. If $h \in H$, and $h = \psi \vee \theta$ $(\psi \wedge \theta)$, then $h \frown \psi \in H, h \frown \theta \in H$. In addition, $P(h) = \exists \ (\forall)$ (we assume that φ is in negation normal form).

4. Every history in H is formed according to the rules (1), (2) and (3).

Obviously, every terminal history $h \in H$ has the form $h = (\varphi_0 \ldots \varphi_{n-1})$, where $\varphi_0 = \varphi$, and φ_{n-1} is an atomic subformula or the negation of an atomic subformula of φ.

- For every terminal history $h = (\varphi_0 \ldots \varphi_{n-1})$:
  - If $M \models \varphi_{n-1}$, then $u_\exists(h) = 1$ and $u_\forall(h) = -1$.
  - If not $M \models \varphi_{n-1}$, then $u_\exists(h) = -1$ and $u_\forall(h) = 1$.

The above rules say in a precise way what could be described informally so that a disjunction (conjunction) prompts a move by player $\exists$ ($\forall$) who chooses the left or the right disjunct. The game terminates with an atomic sentence or its negation. If that sentence is true, then $\exists$ wins; otherwise $\forall$ wins.

The notion of strategy is defined in the same way as in the preceding section.

Example 2.1 *The extensive semantical game $\mathcal{G}((p \vee q) \wedge (q \vee p)), M)$, with $M = \{p\}$ is given in Figure 7. The winning strategy of player $\exists$ is*
$$f_\exists(p \vee q) = p, f_\exists(q \vee p) = p.$$

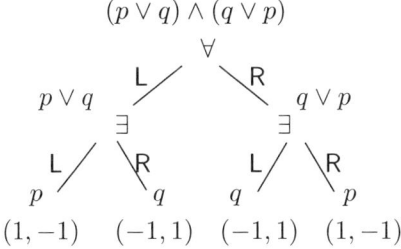

Figure 7: The extensive semantical game $\mathcal{G}((p \vee q) \wedge (q \vee p)), M)$, with $M = \{p\}$

2.2 Semantical games of imperfect information

Semantical games of imperfect information arise in connection with independent connectives, that is, with sentential expressions like

$$(\varphi \ (\vee/\wedge) \ \psi) \wedge (\theta \ (\vee/\wedge) \ \chi) \,,$$

and the games associated with them, played on a model M. The idea here is that these games are supposed to be the analogies to the semantical games of perfect information, except that, after *Abelard* choosing one of the conjuncts of the above expression, *Eloise* does not know which conjunct has been chosen. Then, given the uniformity of the strategies of the players, *Eloise* would have a winning strategy in the game if and only if φ and θ are true in M or ψ and χ are true in M; and *Abelard* would have a winning strategy in the game if and only if φ and ψ are false in M, or ψ and χ are false in M.

The situation in this case is not, however, as smooth as in the case of semantical games of perfect information, for reasons pointed out in the introduction to this paper. More specifically, there is the following problem. Consider the sentence

$$(\varphi_1 \ (\wedge/\vee) \ \varphi_2) \vee p,$$

where conjunction is independent of disjunction. Independence does not seem to make sense in the game-theoretical setting. For after player $\exists$ has made the first move choosing one of the disjuncts, then player $\forall$ must know whether it is his turn to move or not. If it is not his turn to move, then he must be able to infer that $\exists$ has chosen the right-hand disjunct, which being an atomic formula, does not call for a move from any of the players. If again, it *is* his turn to move, then he must know

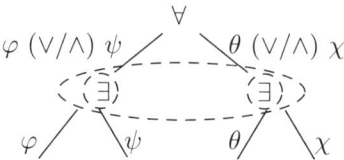

Figure 8: A game for $(\varphi \; (\vee/\wedge) \; \psi) \wedge (\theta \; (\vee/\wedge) \; \chi)$

what move it is that he is supposed to make. And if the move he is supposed to make is to choose one of the conjuncts φ_1 or φ_2, then he must be able to infer that $\exists$ has chosen the left-hand disjunct earlier on. In other words, the fact that $\forall$ knows what moves he is supposed to make at a particular move implies that he knows what stage the game has reached, from which he must be able to infer all the choices $\exists$ has made earlier on. But then the effect of the conjunction being independent of disjunction is cancelled!

We think that the problem is a deep one, and it indicates that we cannot formulate semantical games in the way we did in the previous section. By this we mean that if we want our semantical games to be well-defined instances of games as they are understood in the game-theoretical literature, then we should see to it that the earlier constraints on games of imperfect information are obeyed. And it quickly turns out that the somewhat paradoxical situation we ended up with in the game associated with, e.g., $(\varphi_1 \; (\wedge/\vee) \; \varphi_2) \vee p$ is due to some of those conditions being violated.

In order to see this, let us look at the extensive form of the semantical game associated with $(\varphi \; (\vee/\wedge) \; \psi) \wedge (\theta \; (\vee/\wedge) \; \chi)$ played on an arbitrary model M, where the choices of the players are formulated as in the preceding section. The game frame is described in Figure 8.

We notice that the condition of the *consistency of choices with information* is violated. Thus an argument against informationally independent connectives turns out to be a symptom of a much general phenomenon: the violation of a fundamental condition on well-defined extensive games of imperfect information.

It is obvious that the consistency condition is violated as long as the actions of the players consist of choices of *subformulas*. For this reason, we shall require that in the relevant semantical games, the actions of the players come from the set $A = \{\mathsf{L}(eft), \mathsf{R}(ight)\}$. In other words, we shall introduce a distinction between the actions of the players and their consequences. In the game above, we will say that, e.g., after

Abelard's choice of L, the result is the formula $(\varphi \ (\vee/\wedge) \ \psi)$, etc.

We shall also want to enforce the von Neumann & Morgenstern condition on games. That is, we would not want to have formulas like

$$(p \vee q) \vee [(r \ (\vee/\{\wedge, \vee\}) \ s) \wedge (l \ (\vee/\{\wedge, \vee\}) \ t)],$$

because the extensive form of the associated game is seen to violate that condition.

We can ensure that the above conditions hold by selecting a fragment of propositional logic, whose syntax is given by the following rules:

- Atoms and their negations.
- If φ and ψ are formulas, so are $\varphi \vee \psi$ and $\varphi \wedge \psi$.
- If φ, ψ, θ, and χ are formulas, so are $(\varphi \ (\vee/\wedge) \ \psi) \wedge (\theta \ (\vee/\wedge) \ \chi)$ and $(\varphi \ (\wedge/\vee) \ \psi) \vee (\theta \ (\wedge/\vee) \ \chi)$.

We shall present a semantical game $\mathcal{G}(\varphi, M)$ of imperfect information as a zero-sum extensive game $\mathcal{G}_A = \langle N, H, Z, P, L, (\mathcal{I}_i)_{i \in N}, (u_i)_{i \in N} \rangle$ defined in the earlier section, except that we shall have, as a new element, a labelling function $L: H \backslash Z \to \text{Subform}(\varphi)$.

- $A = \{\mathsf{L}, \mathsf{R}\}$.
- L is a labelling function $L: H \backslash Z \to \text{Subform}(\varphi)$.

H and L are defined inductively:

- $() \in H$ and $L(()) = \varphi$.
- If $L(h)$ is $\psi \vee \theta$, $\psi \wedge \theta$, $\psi \ (\vee/\wedge) \ \theta$ or $\psi \ (\wedge/\vee) \ \theta$, then $h \frown \mathsf{L} \in H, h \frown \mathsf{R} \in H, L(h \frown \mathsf{L}) = \psi$, and $L(h \frown \mathsf{R}) = \theta$.
- If $L(h)$ is $\psi \vee \theta$ or $\psi \ (\vee/\wedge) \ \theta$, then $P(h) = \exists$ (we assume that φ is in negation normal form).
- If $L(h)$ is $\psi \wedge \theta$ or $\psi \ (\wedge/\vee) \ \theta$, then $P(h) = \forall$.
- Every history in H is formed according to the rules above.
- If $L(h)$ is $\varphi \vee \psi$, and $L(h \frown \mathsf{L}) = \varphi, L(h \frown \mathsf{R}) = \psi$, and both φ and ψ are slash-free, then $\{h \frown \mathsf{L}\}, \{h \frown \mathsf{R}\} \in \mathcal{I}_\exists$.
- If $L(h)$ is $\varphi \wedge \psi$, and $L(h \frown \mathsf{L}) = \varphi, L(h \frown \mathsf{R}) = \psi$, and both φ and ψ are slash-free, then $\{h \frown \mathsf{L}\}, \{h \frown \mathsf{R}\} \in \mathcal{I}_\forall$.

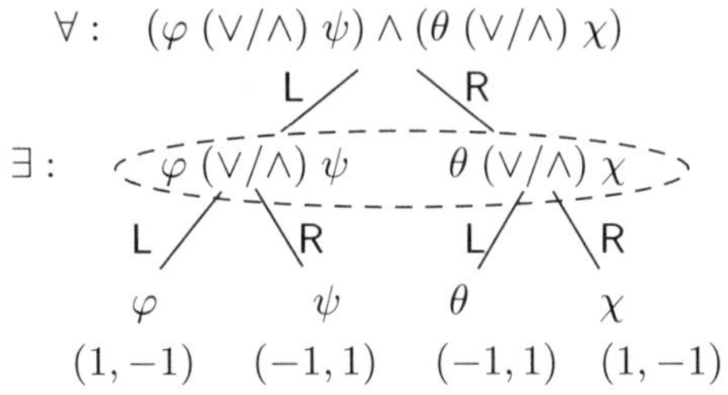

$$\forall: (\varphi \; (\vee/\wedge) \; \psi) \wedge (\theta \; (\vee/\wedge) \; \chi)$$

$$\exists: \quad \varphi \; (\vee/\wedge) \; \psi \qquad \theta \; (\vee/\wedge) \; \chi$$

$$\varphi \qquad \psi \qquad \theta \qquad \chi$$
$$(1,-1) \quad (-1,1) \quad (-1,1) \quad (1,-1)$$

Figure 9: The game $\mathcal{G}((\varphi \; (\vee/\wedge) \; \psi) \vee (\theta \; (\vee/\wedge) \; \chi), M)$, with $M = \{\varphi, \chi\}$.

- If $L(h)$ is $(\varphi \; (\vee/\wedge) \; \psi) \wedge (\theta \; (\vee/\wedge) \; \chi)$, and $L(h \frown \mathsf{L}) = \varphi \; (\vee/\wedge) \; \psi$, and $L(h \frown \mathsf{R}) = \theta \; (\vee/\wedge) \; \chi$, then $\{h \frown \mathsf{L}, h \frown \mathsf{R}\} \in \mathcal{I}_\exists$.

- If $L(h)$ is $(\varphi \; (\wedge/\vee) \; \psi) \vee (\theta \; (\wedge/\vee) \; \chi)$, and $L(h \frown \mathsf{L}) = \varphi \; (\wedge/\vee) \; \psi$, and $L(h \frown \mathsf{R}) = \theta \; (\wedge/\vee) \; \chi$, then $\{h \frown \mathsf{L}, h \frown \mathsf{R}\} \in \mathcal{I}_\forall$.

Obviously, every terminal history $h \in H$ has the form $h = (h_0 \ldots h_{n-1})$, where $L(h_0) = \varphi$, and $L(h_{n-1})$ is an atomic subformula or the negation of an atomic subformula of φ.

- For every terminal history $h = (h_0 \ldots h_{n-1})$:
  - If $M \models L(h_{n-1})$, then $u_\exists(h) = 1$ and $u_\forall(h) = -1$.
  - If not $M \models L(h_{n-1})$, then $u_\exists(h) = -1$ and $u_\forall(h) = 1$.

Example 2.2 *In Figure 9 we have the extensive form of the game $\mathcal{G}((\varphi \; (\vee/\wedge) \; \psi) \vee (\theta \; (\vee/\wedge) \; \chi), M)$, with $M = \{\varphi, \chi\}$. The game structure is similar to the password game of imperfect information discussed earlier. In this case $\mathcal{I}_\exists = \{\{\mathsf{L}, \mathsf{R}\}\}$; $\mathcal{I}_\forall = \{\{()\}\}$. By the requirement on the uniformity of strategies, we have:*

- $M \models^+_{\mathrm{GTS}} (\varphi \; (\vee/\wedge) \; \psi) \wedge (\theta \; (\vee/\wedge) \; \chi) \iff (M \models \varphi \text{ and } M \models \chi)$

- $M \models^-_{\mathrm{GTS}} (\varphi \; (\vee/\wedge) \; \psi) \wedge (\theta \; (\vee/\wedge) \; \chi) \iff (M \models \neg\varphi \text{ and } M \models \neg\chi)$.

Neither Eloise nor Abelard has a winning strategy in this game.

We can extend informationally independent propositional languages with even more complex independent connectives. For instance, adding a clause like

- If $\varphi_1 \ldots \varphi_8$ are sentences, so is

$$[(\varphi_1 \; (\vee/\wedge) \; \varphi_2) \vee (\varphi_3 \; (\vee/\wedge) \; \varphi_4)] \wedge$$
$$[(\varphi_5 \; (\vee/\wedge) \; \varphi_6) \vee (\varphi_7 \; (\vee/\wedge) \; \varphi_8)] \; ,$$

will allow us to have the phenomenon of signalling, exactly as in the extended password game discussed earlier. That is done by the semantical game

$$\mathcal{G}([(\varphi_1 \; (\vee/\wedge) \; \varphi_2) \vee (\varphi_3 \; (\vee/\wedge) \; \varphi_4)] \wedge$$
$$[(\varphi_5 \; (\vee/\wedge) \; \varphi_6) \vee (\varphi_7 \; (\vee/\wedge) \; \varphi_8)], M) \; ,$$

with, say, $M = \{\varphi_1, \varphi_3, \varphi_5, \varphi_7\}$.

2.3 Informationally independent connectives and knowledge of the game

Let us look at semantical games of imperfect information as Kripke structures. The conversion of an extensive zero-sum semantical game $\mathcal{G}(\varphi, M) = \langle \{\exists, \forall\}, H, Z, P, L, (u_i)_{i \in \{\exists, \forall\}} \rangle$ into a Kripke frame $K_{\mathcal{G}} = \langle W, R_\exists, R_\forall \rangle$ was described earlier. According to it we have:

- $W = H$;

- $R_i = \mathcal{I}_i \cup \{\{h\} \mid h \in H, P(h) \neq i\}$.

Applied to the particular example of the password game $\mathcal{G}((p \; (\vee/\wedge) \; q) \vee (q \; (\vee/\wedge) \; p), M)$, with $M = \{p\}$, we get:

- $W = \{(), \mathsf{L}, \mathsf{R}, (\mathsf{L}, \mathsf{L}), (\mathsf{L}, \mathsf{R}), (\mathsf{R}, \mathsf{L}), (\mathsf{R}, \mathsf{R})\}$;

- $R_\forall = \{\{()\}, \{\mathsf{L}\}, \{\mathsf{R}\}, \{(\mathsf{L}, \mathsf{L})\}, \{(\mathsf{L}, \mathsf{R})\}, \{(\mathsf{R}, \mathsf{L})\}, \{(\mathsf{R}, \mathsf{R})\}\}$.

- $R_\exists = \{\{()\}, \{\mathsf{L}, \mathsf{R}\}, \{(\mathsf{L}, \mathsf{L})\}, \{(\mathsf{L}, \mathsf{R})\}, \{(\mathsf{R}, \mathsf{L})\}, \{(\mathsf{R}, \mathsf{R})\}\}$.

Suppose we have in the language the following atomic sentences:

1. **Choice**(L) is an atomic sentence which says: "L is a choice for one of the players";

2. **Choice**(R) is an atomic sentence which says: "R is a choice for one of the players".

This is ensured by a valuation function V such that
$$V(\mathbf{Choice}(\mathsf{L})) = V(\mathbf{Choice}(\mathsf{R})) = \{(), \mathsf{L}, \mathsf{R}\}.$$
Now in the Kripke structure $K_{\mathcal{G}} = \langle W, R_\exists, R_\forall, V \rangle$ we have:

$(S_F, \mathsf{L}) \models K_\exists \mathbf{Choice}(\mathsf{L})$, because $\mathsf{L}, \mathsf{R} \in V(\mathbf{Choice}(\mathsf{L}))$
$(S_F, \mathsf{R}) \models K_\exists \mathbf{Choice}(\mathsf{L})$, because $\mathsf{L}, \mathsf{R} \in V(\mathbf{Choice}(\mathsf{L}))$
$(S_F, \mathsf{L}) \models K_\exists \mathbf{Choice}(\mathsf{R})$, because $\mathsf{L}, \mathsf{R} \in V(\mathbf{Choice}(\mathsf{R}))$
$(S_F, \mathsf{R}) \models K_\exists \mathbf{Choice}(\mathsf{R})$, because $\mathsf{L}, \mathsf{R} \in V(\mathbf{Choice}(\mathsf{R}))$,

and the same holds for *Abelard*. In other words, the requirement that the players should know their choices is satisfied.

There are, however, other death traps which come in from the back door.

Let us add to our language the following atomic sentences:

1. **Label**(φ): a history is labelled with φ;

2. **Label**$((\varphi\ (\vee/\wedge)\ \psi))$: a history is labelled with $(\varphi\ (\vee/\wedge)\ \psi)$;

3. **Label**$((\theta\ (\vee/\wedge)\ \chi))$: a history is labelled with $(\theta\ (\vee/\wedge)\ \chi)$.

That is, we shall set:

$V(\mathbf{Label}(\varphi)) = \{(\mathsf{L}, \mathsf{L})\};$
$V(\mathbf{Label}((\varphi\ (\vee/\wedge)\ \psi))) = \{\mathsf{L}\};$
$V(\mathbf{Label}(\theta\ (\vee/\wedge)\ \chi)) = \{\mathsf{R}\}.$

Then the following statements hold:

$(S_F, \mathsf{L}) \models \neg K_\exists \mathbf{Label}((\varphi\ (\vee/\wedge)\ \psi)) \wedge \neg K_\exists \mathbf{Label}((\theta\ (\vee/\wedge)\ \chi))$
$(S_F, \mathsf{R}) \models \neg K_\exists \mathbf{Label}((\varphi\ (\vee/\wedge)\ \psi)) \wedge \neg K_\exists \mathbf{Label}((\theta\ (\vee/\wedge)\ \chi))$.

In other words, the players act blindly in the sense of not knowing the consequences of their actions! If we require the players to know the labelling formulas in the game, then we have to give up informational independence. So the problems pointed out in Hodges 1997 reappear in a different form. If we enlarge the consistency requirement to cover also the labelling formulas in the extensive form of a semantical game, then we have to give up informational independence.

A way to have both the consistency requirement and informational independence is to treat connectives as *restricted quantifiers*. In this new setting, we shall no longer have formulas like $((\varphi\ (\vee/\wedge)\ \psi) \wedge (\theta\ (\vee/\wedge)\ \chi))$, but instead formulas of the form

$$\forall i_1 (\exists i_2 / \{i_1\})\, p_{i_1 i_2},$$

where i_1 and i_2 range over the domain $\{1, 2\}$, and in addition:

$$p_{11} = \varphi, p_{12} = \psi, p_{21} = \theta, p_{22} = \chi.$$

One can now check that all the conditions on games of imperfect information discussed earlier are satisfied (see Sandu and Pietarinen 2001).

References

van Benthem, J. 2001. Hintikka self-applied. In Lewis, H., editor, *Library of Living Philosophers: Jaakko Hintikka*. To Appear. Available electronically as http://staff.science.uva.nl/~johan/H-H.ps.

Bonanno, G. and P. Battigalli. 1997. Synchronic information, knowledge and common knowledge in extensive games. In *Epistemic Logic and the Theory of Games and Decisions*, pages 235–264. Kluwer, Dordrecht.

Henkin, L. 1961. Some remarks on infinitely long formulas. In *Infinistic Methods. Proceedings of the Symposium on Foundations of Mathematics, Warsaw, Panstwowe, 2–9 September 1959*, pages 167–183, New York. Pergamon Press. (no editor given).

Hintikka, J. and G. Sandu. 1997. Game-theoretical semantics. In van Benthem, J. and A. ter Meulen, editors, *Handbook of Logic and Language*, pages 361–410. Elsevier, Amsterdam.

Hodges, W. 1997. Compositional semantics for a language of imperfect information. *Logic Journal of the IGPL*, 5:539–563.

Osborne, M. and A. Rubinstein. 1994. *A Course in Game Theory*. MIT Press, Cambridge, MA.

Rubinstein, A. 1998. *Modeling Bounded Rationality*. MIT Press, Cambridge, MA.

Sandu, G. and A. Pietarinen. 2001. Partiality and games: Propositional logic. *Logic Journal of the IGPL*, 9:107–127.

Descriptions of Game States

Hans van Ditmarsch, Wiebe van der Hoek, Barteld Kooi

> Knowledge games are a fair playground to study properties of multiagent systems. In the initial state of a knowledge game a number of cards is dealt over a number of players. We describe this initial game state by a theory comprising facts about deals of cards and the agents' knowledge and ignorance of those facts. We also provide a Kripke ($S5$) model for such a game state and show that it is unique up to bisimulation. Therefore this model demonstrates all and only the relevant epistemic properties of the initial state.

1 Introduction

The interaction between logic and game theory is currently of interest to the scientific community. Well-known are game theoretical foundations for logical semantics (Hintikka and Sandu, 1989), and other applications of game theory in logic (Ehrenfeucht, 1961; van Benthem, 2000a). For applications of logic in game theory, we may mention the formalization in logical theories of game theoretical notions such as game trees, plays of a game, and equilibria (Bonanno, 1993; Kaneko and Nagashima, 1996). One issue of interest in this area are games where the information contained in a game state and the information change due to a game action may be rather complex, and therefore become objects of study in themselves. As a concrete example of such games we have defined knowledge games (van Ditmarsch, 2001): card games where a number of cards is distributed over a number of players (in the context of games, we prefer to call agents 'players'), and where moves consist of information exchange, such as showing cards to other

---

*Addresses: Hans van Ditmarsch, Computer Science, University of Otago, Dunedin, New Zealand, hans@cs.otago.ac.nz. Wiebe van der Hoek, Computer Science, Utrecht University, The Netherlands, and University of Liverpool, UK, wiebe@cs.uu.nl. Barteld Kooi, Department of Mathematics and Computing Science, University of Groningen, The Netherlands, barteld@cs.rug.nl.

The authors thank the anonymous referees for their useful comments. Hans van Ditmarsch and Barteld Kooi thank CSLI for its hospitality.

Games, Logic, and Constructive Sets
Grigori Mints and Reinhard Muskens (eds.)
Copyright ©2003, CSLI Publications

players. Of particular interest are actions where information is simultaneously exchanged between subgroups of different size, e.g. when one player shows a card to another player, while the remaining players see that a card is being shown but not which card it is.

From a given knowledge game state and a game action that is executable in that state we can compute the next game state (van Ditmarsch, 2000). Therefore, we can compute any game state from an initial knowledge game state and a game action sequence. This illustrates the need for a logical description of initial game states. Although it seems to be rather clear what the players know about *facts* in an initial game state, the precise amount of their ignorance of *other players' knowledge* is less transparent, as one easily overlooks game features. It is important to know what other players know, because in other game states this may help one to derive factual knowledge, which may help to win.

In this article we provide descriptions of initial knowledge game *states*. For the description of game *actions*, see (van Ditmarsch, 2000).

In section 2 we give an overview of relevant logical terminology. Readers familiar with epistemic logic may prefer to skip this section. See also (Meyer and van der Hoek, 1995; Fagin et al., 1995). In section 3 we present the theory **Hexa** that describes the model $Hexa$ of the initial game state of the knowledge game for three players each holding one card. This knowledge game exemplifies most of the features that we want to study. In section 4 we continue with the general case: the theory **Kgames** describes the initial game state of the knowledge game for an arbitrary deal of cards over players. In section 5 we describe the game state where the cards have been dealt but where players haven't picked up and looked at their cards.

2 Epistemic logic

Given (throughout the text) are finite sets A of agents and P of atoms.

Syntax of epistemic logic Multiagent epistemic logic $\mathcal{L}_A^P$ is the smallest set such that, if $p \in P, \varphi, \psi \in \mathcal{L}_A^P, a \in A, B \subseteq A$, then: $p, \neg\varphi, (\varphi \wedge \psi), K_a\varphi, C_B\varphi \in \mathcal{L}_A^P$. Formula $K_a\varphi$ means '*a knows φ*'. Formula $C_B\varphi$ means '*B know φ*' (or '*B commonly know φ*'). If $B = A$ we also say that '*B publicly know φ*'. We introduce the usual abbreviations $\varphi \vee \psi$, $\varphi \rightarrow \psi$, and $\varphi \leftrightarrow \psi$.

Epistemic structures An $S5_A$ *model* is a triple $\langle W, \{\sim_a\}_{a\in A}, V\rangle$ where W is the (nonempty) *domain*, for each $a \in A$, $\sim_a \subseteq W \times W$ is the *accessibility relation* for agent a, which is an *equivalence relation*, and the *valuation* V is a function that given a world assigns a truth value to each atom: $V : W \to P \to \{0,1\}$ (i.e. for each $w \in W$, $V_w : P \to \{0,1\}$). Instead of $w \in W$ we also write $w \in M$. The pair $\langle W, \{\sim_a\}_{a\in A}\rangle$ is an $S5_A$ *frame*. If $w \in M$, the pair (M, w) is an $S5_A$ *state*, where w is its *point* or *designated world*.

Define $\sim_B := (\bigcup_{a\in B} \sim_a)^*$ (* stands for transitive and reflexive closure). An $S5_A^C$ model is an $S5_A$ model with accessibility relations $\sim_B$ added for all $B \subseteq A$. Every $S5_A^C$ model can also be seen as an $S5_A$ model. We generally write $S5$ instead of $S5_A$, and we always assume an arbitrary model to be an $S5$ model.

Semantics of epistemic logic Let $M = \langle W, \{\sim_a\}_{a\in A}, V\rangle$ be an $S5$ model, $w \in M$, $p \in P$, $a \in A$, $B \subseteq A$. Then: $M, w \models p \Leftrightarrow V_w(p) = 1$; $M, w \models \neg\varphi \Leftrightarrow M, w \not\models \varphi$; $M, w \models \varphi \wedge \psi \Leftrightarrow [\ M, w \models \varphi$ and $M, w \models \psi\ ]$; $M, w \models K_a\varphi \Leftrightarrow \forall w' \sim_a w : M, w' \models \varphi$; $M, w \models C_B\varphi \Leftrightarrow \forall w' \sim_B w : M, w' \models \varphi$. Derived notions are: $M \models \varphi \Leftrightarrow \forall w \in M : M, w \models \varphi$ (note that we have $M \models \varphi \Leftrightarrow M \models C_A\varphi$), and $M, w \models \Sigma \Leftrightarrow \forall \varphi \in \Sigma : M, w \models \varphi$.

We now define (local) logical consequence. We write $\models$ instead of, strictly, $\models_{S5_A}$. We define $\Sigma \models \varphi \Leftrightarrow \forall M, \forall w \in M : M, w \models \Sigma \Rightarrow M, w \models \varphi$.

Bisimulation Let $M = \langle W, \{\sim_a\}_{a\in A}, V\rangle$, $N = \langle W'', \{\sim_a''\}_{a\in A}, V''\rangle$ be $S5$ models. A *bisimulation* between two states (M, x) and (N, y) is a nonempty binary relation $\mathfrak{R} \subseteq W \times W''$ such that $\mathfrak{R}(x, y)$ and:

Atoms: $\forall w \in M, \forall v \in N : \mathfrak{R}(w, v) \Rightarrow V_w = V''_v$;

Forth: $\forall a \in A, \forall w, w' \in M, \forall v \in N : \mathfrak{R}(w, v)$ and $w \sim_a w' \Rightarrow \exists v' \in N : v \sim_a'' v'$ and $\mathfrak{R}(w', v')$;

Back: $\forall a \in A, \forall w \in M, \forall v, v' \in N : \mathfrak{R}(w, v)$ and $v \sim_a'' v' \Rightarrow \exists w' \in M : w \sim_a w'$ and $\mathfrak{R}(w', v')$.

We say that (M, x) *is bisimilar to* (N, y) and we write $(M, x) \underline{\leftrightarrow} (N, y)$. If $\mathsf{domain}(\mathfrak{R}) = M$ and $\mathsf{range}(\mathfrak{R}) = N$, M *is bisimilar to* N, and we write $M \underline{\leftrightarrow} N$. It holds that $(M, w) \underline{\leftrightarrow} (N, v) \Rightarrow [\ \forall \varphi : M, w \models \varphi \Leftrightarrow N, v \models \varphi\ ]$, and that $M \underline{\leftrightarrow} N \Rightarrow [\ \forall \varphi : M \models \varphi \Leftrightarrow N \models \varphi\ ]$.

Description Let M be an $S5$ model, $w \in M$. The *atomic description* φ of world w (or of state M, w) is $\bigwedge_{p \in P} p^w$, where $p^w := p$ if $V_w(p) = 1$ and $p^w := \neg p$ if $V_w(p) = 0$. A *description of a model* M is a set of formulas (i.e. a *theory*) Σ such that $M \models \Sigma$ and $\forall M' : M' \models \Sigma \Rightarrow M' \leftrightarrow M$. In other words: M is a model of Σ, and *only* M is a model of Σ. A *description of a state* (M, w) is a set Σ such that $M, w \models \Sigma$ and $\forall M', \forall w' \in M' : M', w' \models \Sigma \Rightarrow (M, w) \leftrightarrow (M', w')$.[1] Finally, a *characteristic* φ of M is merely a formula that is valid in M: $\forall w \in M : M, w \models \varphi$.

3 Description of Hexa

In this section, we present the theory Hexa, describing the $(S5_{\{1,2,3\}})$ model $Hexa$. The state $(Hexa, rwb)$ is the initial state of the knowledge game for three players each holding a card, where 1 holds red (r), 2 holds white (w), and 3 holds blue (b). In such games, moves consist of questions, such as "Do you have the white or the blue card?", and their answers, such as "No". We disregard all game aspects and merely focus on the logical description of the initial game state. Figure 1 pictures the model $Hexa$. In the figure, a node labelled cde represents the deal of cards where 1 holds c, 2 holds d and 3 holds e; an arc is labelled with a if player a cannot distinguish from each other the deals that are linked by it.

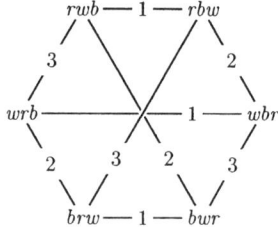

Figure 1: The model $Hexa$ for three players each holding a card

What information do the players have in this model, regardless of the actual deal of cards? They know how many cards there are, namely three. They know that the cards are all different, namely one red, one white and one blue. They know that each of them holds one card. Beyond that, if they hold a card, they know it, and if they don't hold

---
[1] Occasionally we also use 'describe' in an informal sense, as in 'atom c_a describes that player a holds card c'.

a card, they also know that they do not hold it. All this is publicly known. They don't know anything else, and there seem to be two sides of that ignorance. First, a player doesn't know that another player holds a specific card. Second, with the exception of his own card, a player can imagine any card to be in possession of another player. All these constraints are satisfied by the theory Hexa.

Definition 1 (Hexa) *Theory* Hexa *consists of the formulas:*

$$\begin{aligned}
\text{seeH} &:= \bigwedge\nolimits_{a\in\{1,2,3\}} \bigwedge\nolimits_{c\in\{r,w,b\}} (c_a \to K_a c_a) \\
\text{dealsH} &:= \delta_{rwb} \vee \delta_{rbw} \vee \delta_{wrb} \vee \delta_{wbr} \vee \delta_{brw} \vee \delta_{bwr} \\
\text{dontknowthatH} &:= \bigwedge\nolimits_{a\neq b\in\{1,2,3\}} \bigwedge\nolimits_{c\in\{r,w,b\}} \neg K_a c_b
\end{aligned}$$

These formulas are the *constituents* of the theory. Atomic proposition c_a expresses that player a holds card c. SeeH expresses that every agent can see his own card. DealsH expresses that there are only six relevant deals, δ_{abc} is the atomic description of the world (deal) abc in $Hexa$, i.e. $a_1 \wedge \neg b_1 \wedge \neg c_1 \wedge \neg a_2 \wedge b_2 \wedge \neg c_2 \wedge \neg a_3 \wedge \neg b_3 \wedge c_3$. DontknowthatH expresses that players do not know the cards of other players. In section 3.1 we discuss various properties of agent knowledge in $Hexa$ that can be derived from the concise formulation in Hexa.

Theorem 1 *Theory* Hexa *describes model* $Hexa$.

We have to show that $Hexa \models$ Hexa, which is obvious, and that only $Hexa$ models Hexa. The last follows from the following lemma, that is a special case of lemma 6, to be presented and proven in section 4:

Lemma 2 *Suppose* $M \models$ Hexa. *Then* $M \leftrightarrow Hexa$.

What does it mean that Hexa describes $Hexa$? Only in a technical sense is $Hexa$ the only model of Hexa: Hexa defines the bisimulation class of $Hexa$. But from this it follows that every non-theorem of Hexa is falsified in $Hexa$, and every theorem of Hexa is valid in $Hexa$. Differently said, if we *weaken* theory Hexa, e.g. by deleting formulas, it models structures of unintended game states, but if we *strengthen* the theory, it no longer models $Hexa$.

When we say that Hexa describes $Hexa$, we have implicitly quantified over all its worlds. This has to be made explicit when we describe one of its states: we now demand that Hexa is common knowledge in that state, and that its (unique) atomic description holds:

Corollary 3 *Theory* C_{123}Hexa $+ \delta_d$ *describes* $(Hexa, d)$.[2]

---
[2] Where C_{123}Hexa $+ \delta_d = \{C_{123}\varphi \mid \varphi \in$ Hexa$\} \cup \{\delta_d\}$.

Corollary 4 *For all $\mathcal{L}_{123}^{\{r_1,r_2,\ldots\}}$ formulas φ and ψ: Hexa $\models \varphi \Leftrightarrow$ Hexa $\models C_{123}\varphi \Leftrightarrow C_{123}$Hexa $\models \varphi \Leftrightarrow C_{123}$Hexa $\models C_{123}\varphi$ and Hexa, $d \models \psi \Leftrightarrow C_{123}$Hexa $+ \delta_d \models \psi$.*

Corollary 4 is a direct consequence of theorem 1. Note that Hexa $\models \varphi \not\Leftrightarrow$ Hexa $\models C_{123}\varphi$, as e.g. deals $\not\models C_{123}$deals.

In a way, theorem 1 also shows that we have chosen the *right* model, and the *right* description, for the game state of three players each holding a card. From the viewpoint of knowledge specification, we were initially uncertain about *both* the properties of the agents in this game state *and* the $S5$ model representing it. But by linking these so tightly together in theorem 1 we have simultaneously validated them, so to speak.

3.1 Derived characteristics of Hexa

One can define various other characteristics of $Hexa$. We list a few:

Definition 2 (Other properties of players' knowledge)

$$
\begin{aligned}
&\text{players only see their own cards}\\
&\text{dontseeH} := \bigwedge_{a\in\{1,2,3\}} \bigwedge_{c\in\{r,w,b\}} (\neg c_a \to K_a \neg c_a)\\
&\text{there is at most one card of each colour}\\
&\text{atmostH} := \bigwedge_{a\neq b\in\{1,2,3\}} \bigwedge_{c\in\{r,w,b\}} \neg(c_a \wedge c_b)\\
&\text{there is at least one card per player}\\
&\text{atleastH} := \bigwedge_{a\in\{1,2,3\}} (r_a \vee w_a \vee b_a)\\
&\text{players can imagine that others hold other cards}\\
&\text{dontknownotH} := \bigwedge_{a\neq b\in\{1,2,3\}} \bigwedge_{c\in\{r,w,b\}} (\neg c_a \to \neg K_a \neg c_b)\\
&\text{players do not know that others hold other cards}\\
&\text{dontknowotherH} := \bigwedge_{a\neq b\in\{1,2,3\}} \bigwedge_{c\in\{r,w,b\}} (\neg c_a \to \neg K_a c_b)
\end{aligned}
$$

In dontknownotH, we may read $\neg K_a \neg$ as the dual epistemic modal operator meaning 'player a can imagine that' (M_a). All these properties hold in $Hexa$, and by corollary 4 therefore follow from C_{123}Hexa as well. Something more than that can be observed, however. We give an example: although seeH and dontseeH are not logically equivalent, we can replace seeH by dontseeH in theory Hexa and the result will *still* describe $Hexa$. This is captured by the following notion:

Definition 3 (Stronger) *Let Σ, φ, ψ be $\mathcal{L}_A^P$ formulas. Then 'φ is stronger than ψ given Σ', notation $\varphi \geq_\Sigma \psi$, if $C_A(\Sigma - \psi + \varphi) \models \psi$.*[3]

---

[3] $\Sigma - \psi = \Sigma \setminus \{\psi\}$ if $\psi \in \Sigma$ and $\Sigma - \psi = \Sigma$ otherwise. $\Sigma + \varphi = \varphi + \Sigma = \Sigma \cup \{\varphi\}$; whether φ is already in Σ or not makes no difference.

If either $\varphi \in \Sigma$ or $\psi \in \Sigma$, we say that 'φ is stronger than ψ in Σ'. If $\varphi \geq_\Sigma \psi$ and $\psi \geq_\Sigma \varphi$ we write $\varphi =_\Sigma \psi$, and we say that 'φ and ψ are equally strong given/in Σ' (or 'φ is just as strong as ψ'). One can show that: dealsH $=_{\text{Hexa}}$ atmostH $\wedge$ atleastH, that seeH $=_{\text{Hexa}}$ dontseeH, that dontknowthatH $=_{\text{Hexa}}$ dontknowotherH, and that dontknowthatH $=_{\text{Hexa}}$ dontknownotH. Although the proofs need some combinatorial juggling, they are rather basic and have been omitted.

4 Description of initial game states

In the previous section we have described *Hexa*. In this section we generalize our results for any number of players and cards. We start by introducing terminology on deals of cards.

We consider deals $d \in A^\mathbf{C}$ of a set $\mathbf{C}$ of cards (in bold, to distinguish it from the common knowledge operator) over a finite set A of players. The cards held by player a can be represented by $d^{-1}(a) = \{c \in \mathbf{C} \mid d(c) = a\}$. Consider the game state where the cards have been dealt and every player has (only) looked at his own cards. In this *initial game state* two deals can not be distinguished from each other by a player (i.e., are the same for a player) if he holds the same cards in both *and* if all other players hold the same number of cards in both (imagine counting the backfaces of other players' cards). This induces an equivalence on card deals: *deal d is the same for player a as deal e* iff [$d^{-1}(a) = e^{-1}(a)$ and $\forall b \in A : |d^{-1}(b)| = |e^{-1}(b)|$]. We introduce some useful abbreviations. Let $a, b \in A$, $d \in A^\mathbf{C}$. Then $\sharp a := |d^{-1}(a)|$ is the number of cards held by player a. Similarly, write $\sharp \neg a$ for $|\mathbf{C}| - |d^{-1}(a)|$, the number of cards *not* held by a, and write $\sharp \neg ab$ for $|\mathbf{C}| - |d^{-1}(a)| - |d^{-1}(b)|$, the number of cards not held by a or b. The *size* of deal d, notation $\sharp d$, lists for each player the number of cards he holds.

In the following definition we represent the initial game state by an $S5$ model. The worlds in our model are deals. The atoms P of our language now take the form c_a, for $a \in A$ and $c \in \mathbf{C}$.

Definition 4 (Initial game state) *The initial game state is modelled by the pointed $S5_A$ model $(I_d, d) = (\langle D_{\sharp d}, \{\sim_a\}_{a \in A}, V\rangle, d)$, where $D_{\sharp d} = \{e \in A^\mathbf{C} \mid \forall a \in A : |e^{-1}(a)| = |d^{-1}(a)|\}$, and for all $a \in A$, $c \in \mathbf{C}$, $d_1, d_2 \in D_{\sharp d}$: $d_1 \sim_a d_2 \Leftrightarrow d_1^{-1}(a) = d_2^{-1}(a)$, and $V_{d_1}(c_a) = 1 \Leftrightarrow d_1(c) = a$.*

Definition 5 (Description of a deal of cards) *Let $d \in A^\mathbf{C}$. The description of deal d is the atomic description δ_d (see section 2) of the* world d in I_d: $\delta_d := \bigwedge_{a \in A} \bigwedge_{c \in \mathbf{C}} c_a^d$. The description of the cards of player a is the part of that description about a: $\delta_d^a := \bigwedge_{c \in \mathbf{C}} c_a^d$.

By definition we have that $\delta_d \leftrightarrow \bigwedge_{a \in A} \delta_d^a$. Also, for arbitrary $d', d'' \in I_d$ and $a \in A$: $\delta_d^a \leftrightarrow \bigvee_{d' \sim_a d} \delta_{d'}$, and $d \sim_a d' \Rightarrow \delta_d^a \leftrightarrow \delta_{d'}^a$. For more details on card deals and knowledge games, see (van Ditmarsch, 2001). We now present the theory **Kgames**:

Definition 6 (Kgames) *The theory* **Kgames** *for deal* $d \in A^{\mathbf{C}}$ *consists of the three constituents:*

$$\begin{aligned}
\text{deals} &:= \bigvee_{d' \in D_{\sharp d}} \delta_{d'} \\
\text{seedeal} &:= \bigwedge_{a \in A} \bigwedge_{d' \in D_{\sharp d}} (\delta_{d'}^a \to K_a \delta_{d'}^a) \\
\text{dontknow} &:= \bigwedge_{a \in A} \bigwedge_{d' \in D_{\sharp d}} (\delta_{d'}^a \leftrightarrow \neg K_a \neg \delta_{d'})
\end{aligned}$$

Deals expresses that every world is atomically characterized by a deal of size $\sharp d$. Seedeal expresses that you know your own cards. Dontknow expresses that you can imagine any deal that is consistent with (in the sense of 'extends') your own cards. In a way, **deals** expresses the *factual knowledge* of all players, **seedeal** expresses the *private knowledge* of a player, and **dontknow** expresses his *private ignorance*. Again, all this is *publicly* known. In the coming subsections we discuss various other characterizations of the players' knowledge in I_d, and in what respect they are generalizations of constituents of **Hexa**. First, we show that **Kgames** describes I_d.

Theorem 5 *Theory* **Kgames** *for deal* $d \in A^{\mathbf{C}}$ *describes* I_d.

We have to show that I_d is a model of **Kgames**, which is obvious, and that *only* I_d models **Kgames**:

Lemma 6 *Suppose* $M \models$ **Kgames**. *Then* $M \underline{\leftrightarrow} I_d$.

Proof Let $M = \langle W^M, \{\sim_a^M\}_{a \in A}, V^M \rangle$ and $M \models$ **Kgames**; $I_d = \langle D_{\sharp d}, \{\sim_a\}_{a \in A}, V \rangle$. Observe that, because $M \models$ **deals**, each world $w \in M$ has a valuation $V_w = V_{d'}$ for some $d' \in D_{\sharp d}$. Define relation $\mathfrak{R} \subseteq (W^M \times D_{\sharp d})$ as follows: $\forall w \in W^M : \forall d' \in D_{\sharp d} : \mathfrak{R}(w, d') \Leftrightarrow V_w = V_{d'}$. We prove that $\mathfrak{R}$ is a bisimulation between M and I_d.

Atoms: By definition of $\mathfrak{R}$.

Forth: Let $w, w' \in M$, let $e \in D_{\sharp d}$. Suppose that $\mathfrak{R}(w, e)$ and that, for an arbitrary $a \in A$: $w \sim_a w'$. We find an $\mathfrak{R}$-image in $D_{\sharp d}$ of w' as follows:

Observe that $I_d, e \models \delta_e$. As $V_w = V_e$, also $M, w \models \delta_e$. Therefore $M, w \models \neg K_a \neg \delta_e$. From that and $M, w \models$ **dontknow** follows $M, w \models \delta_e^a$, i.e.: $M, w' \models \bigvee_{e' \sim_a e} \delta_{e'}$. Therefore there is an $e' \sim_a e$ such that $M, w' \models \delta_{e'}$. Choose one, then *that* e' is the required $\mathfrak{R}$-image of w': note that $e \sim_a e'$, and that $V_{w'}^M = V_{e'}$, because also, obviously, $I_d, e' \models \delta_{e'}$.

Back: Let $e, e' \in I_d$, let $w \in M$. Suppose that $\mathfrak{R}(w, e)$ and that, for an arbitrary $a \in A$, $e \sim_a e'$. We find an $\mathfrak{R}$-original of e', in M, as follows:

$M, w \models \delta_e$ implies, using **seedeal**, that $M, w \models K_a \delta_e^a$. Let $w'' \sim_a w$ be arbitrary. Then $M, w'' \models \delta_e^a$. From that and $e \sim_a e'$ follows $M, w'' \models \delta_{e'}^a$. Because w'' was arbitrary, $M, w \models K_a \delta_{e'}^a$. Therefore $M, w \models \delta_{e'}^a$. Using **dontknow**, we get $M, w \models \neg K_a \neg \delta_{e'}$. Therefore, there must be a $w' \sim_a w$ such that $M, w' \models \delta_{e'}$. Choose one, then *that* w' is the $\mathfrak{R}$-original of e', as $M, w' \models \delta_{e'}$ expresses that $V_{w'} = V_{e'}$. □

Note that in the 'forth' part of the proof, we have only essentially used the $\neg K_a \neg \delta_e \rightarrow \delta_e^a$ part of **dontknow**, whereas in the 'back' part of the proof, we have essentially used **seedeal** and only used the $\delta_e^a \rightarrow \neg K_a \neg \delta_e$ part of **dontknow**.

Just as in the case of three players and three cards, that **Kgames** describes I_d means that we can neither *weaken* nor *strengthen* the theory, because I_d is its only model.

Corollary 7 *Theory* **Kgames** $+ \delta_e$ *describes* (I_d, e).

Corollary 8 *For all $\varphi, \psi \in \mathcal{L}_A^P$ (where $P = \mathbf{C} \times A$), and $e \in I_d$: $I_d \models \varphi \Leftrightarrow I_d \models C_A \varphi \Leftrightarrow C_A$**Kgames** $\models \varphi \Leftrightarrow C_A$**Kgames** $\models C_A \varphi$, and $I_d, e \models \psi \Leftrightarrow$ **Kgames** $+ \delta_e \models \psi$.*

In corollaries 7 and 8 it does not matter which point e we take in I_d, since $I_d = I_e$ as long as $\sharp d = \sharp e$. In particular, from corollary 7 it follows that the initial state (I_d, d) is described by **Kgames** $+ \delta_d$.

We continue with discussing various derived characteristics of the initial game state.

4.1 Derived characteristics: factual knowledge

Similarly to the case for $Hexa$, there are different ways to express how many cards players hold. As in section 3, we have that **deals** $=_{\text{Kgames}}$ **atmost** $\wedge$ **atleast**.

Definition 7 (How many cards)

$$
\begin{aligned}
\textsf{atmost} &:= \bigwedge_{a \neq b \in A} \bigwedge_{c \in \mathbf{C}} \neg(c_a \wedge c_b) \\
\textsf{atleast} &:= \bigwedge_{a \in A} \bigvee_{c^1 \neq \ldots \neq c^{\sharp a} \in \mathbf{C}} \bigwedge_{i=1}^{\sharp a} c_a^i
\end{aligned}
$$

4.2 Derived characteristics: private knowledge

We list six formulas describing what players 'see', that are all equally strong in Kgames.

Definition 8 (Seeing cards)

$$
\begin{aligned}
\text{see} &:= \bigwedge_{a \in A} \bigwedge_{c \in C} (c_a \to K_a c_a) \\
\text{dontsee} &:= \bigwedge_{a \in A} \bigwedge_{c \in C} (\neg c_a \to K_a \neg c_a) \\
\text{seeall} &:= \bigwedge_{a \in A} \bigwedge_{c^1 \neq \ldots \neq c^{\sharp a} \in C} (\bigwedge_{i=1}^{\sharp a} c_a^i \to K_a \bigwedge_{i=1}^{\sharp a} c_a^i) \\
\text{dontseeall} &:= \bigwedge_{a \in A} \bigwedge_{c^1 \neq \ldots \neq c^{\sharp \neg a} \in C} (\bigwedge_{i=1}^{\sharp \neg a} \neg c_a^i \to K_a \bigwedge_{i=1}^{\sharp \neg a} \neg c_a^i) \\
\text{seedeal} &:= \bigwedge_{a \in A} \bigwedge_{d' \in D_{\sharp d}} (\delta_{d'}^a \to K_a \delta_{d'}^a) \\
\text{dontseedeal} &:= \bigwedge_{a \in A} \bigwedge_{d' \in D_{\sharp d}} (\neg \delta_{d'}^a \to K_a \neg \delta_{d'}^a)
\end{aligned}
$$

See is the generalization of seeH, and dontsee of dontseeH: for every player and for every single card, if a player holds it, he knows that, and if he doesn't hold it, he knows that too. Instead, seeall and dontseeall express that, if a player holds his given number of cards, he knows them all, and, respectively, that if he does not hold any other card, he knows that too. Seedeal expresses that, if a player holds his given number of cards and does not hold any other card, he knows that. This is another way of saying that he knows his *local state* (see also section 6): for a given player a, to every atom c_a or $\neg c_a$ in δ_d^a corresponds the value of a 'local state variable for that player'. Dontseedeal expresses that if a player is *not* in a given local state, he knows that too. All six forms of 'seeing' are equally strong in Kgames. The proofs are simple and use deals. Although see is the most straightforward of all six, we prefer the more abstract formulation in constituent seedeal of Kgames.

4.3 Derived characteristics: private ignorance

The constituent dontknow of Kgames expresses private ignorance. It is conveniently formulated using the notations δ_d for the description of a deal and δ_d^a for the description of the cards of a. Directly reasoning from the properties of agents' knowledge, and starting from the characterization of private ignorance in $Hexa$ by dontknowthatH $= \bigwedge_{a \neq b \in \{1,2,3\}} \bigwedge_{c \in \{r,w,b\}} \neg K_a c_b$, it is not so obvious how to get there. We illustrate this with an example:

Example 1 *Consider the initial game state for the game for three players 1, 2, 3 each holding two cards, with actual deal $kl|mn|op$. How ignorant is player 1 in this state of the game? Clearly, it is not strong enough that 1 does not know that 2 holds a specific combination of two* cards: for all $c \neq c'$, $\neg K_1(c_2 \wedge c'_2)$. Player 1 also does not know that 2 holds any single card, which is stronger: for all c, $\neg K_1 c_2$. However, even that is not strong enough:

Suppose 2 has told the others that he holds one of m and n. After that, it still holds that 1 doesn't know any of 2's cards: $\neg K_1 m_2$ and $\neg K_1 n_2$. However, 1 is less ignorant than before, because he now knows that 2 holds one of two cards: $K_1(m_2 \vee n_2)$. Initially, he didn't know that, indeed: for all $c \neq c' \in \mathbf{C}$: $\neg K_1(c_2 \vee c'_2)$.

It appears that this is exactly the limit of his ignorance, because for some combinations of three cards he does know that 2 holds one of them, e.g. $K_1(m_2 \vee n_2 \vee o_2)$. Suppose not, then it would be conceivable for player 1 that player 2 did not have any of the three cards m, n, o. Because player 1 holds two other cards himself, k, l, there would then be only one card left for player 2 to hold: card p. That player 2 holds only one card, contradicts deals.

Definition 9 (Ignorance) Let $d \in A^{\mathbf{C}}$. Then:

$$\begin{aligned}
\mathsf{dontknowthat} &:= \bigwedge_{a \neq b \in A} \bigwedge_{c^1 \neq \ldots \neq c^{\sharp \neg ab} \in C} \neg K_a \bigvee_{i=1}^{\sharp \neg ab} c_b^i \\
\mathsf{dontknownot} &:= \bigwedge_{a \neq b \in A} \bigwedge_{c^1 \neq \ldots \neq c^{\sharp b} \in C} (\bigwedge_{i=1}^{\sharp b} \neg c_a^i \to \neg K_a \neg \bigwedge_{i=1}^{\sharp b} c_b^i) \\
\mathsf{dontknow} &:= \bigwedge_{a \in A} \bigwedge_{d' \in D_{\sharp d}} (\delta_{d'}^a \leftrightarrow \neg K_a \neg \delta_{d'})
\end{aligned}$$

The generalization of dontknowthatH reached in example 1 expresses characteristic dontknowthat in definition 9. Somewhat similarly, dontknownotH can be generalized to dontknownot. One can show that these two are equally strong in {deals, seedeal}. This is surprising, because dontknownot and dontknowthat appear to describe complementary kinds of ignorance: the first is on ignorance about cards other players have, the second is on ignorance about cards they *don't* have. However, both are still not strong enough to pin down I_d. We present another example to illustrate that:[4]

Example 2 *Consider the initial game state for the game for four players $1, 2, 3, 4$ each holding one card, with actual deal $k|l|m|n$ (klmn). Suppose an outsider tells player 1 that kmln is not the actual deal of cards, so that $K_1 \neg \delta_{kmln}$. Deal kmln is consistent with the local state, k, of player 1 in actual deal klmn, so that $\neg K_1 \neg \delta_{kmln}$ follows from dontknow. Therefore, dontknow does not hold in the current game state. However, dontknowthat holds: for any two cards, 1 can imagine that 2 does not have both. (Similarly, dontknownot holds.)*

---

[4]Suggested by Erik Krabbe.

We summarize the results: dontknowthat $=_{\{\text{deals,seedeal}\}}$ dontknownot and dontknow $\geq_{\text{Kgames}}$ dontknowthat (therefore also dontknow $\geq_{\text{Kgames}}$ dontknownot). We show that dontknownot $\geq_{\{\text{deals,seedeal}\}}$ dontknowthat. For the remaining proofs, see (van Ditmarsch, 2000):

Suppose not [dontknownot $\geq_{\{\text{deals,see}\}}$ dontknowthat]. Then there are players a, b and cards $c^1, ..., c^{\sharp \neg ab}$ such that $K_a \bigvee_{i=1}^{\sharp \neg ab} c_b^i$. Regardless of whether a holds some of the cards $c^1, ..., c^{\sharp \neg ab}$, there must be at least $\sharp b$ other cards that a doesn't hold, suppose: $ca^1, ..., ca^{\sharp b}$. In other words, we have that: $\neg ca_a^1 \wedge ... \wedge \neg ca_a^{\sharp b}$. Applying dontknownot we get $\neg K_a \neg \bigwedge_{i=1}^{\sharp b} ca_b^i$. Formula $\neg K_a \neg \bigwedge_{i=1}^{\sharp b} ca_b^i$ means that a can imagine that b holds the $\sharp b$ cards $ca^1, ..., ca^{\sharp b}$. From $K_a \bigvee_{i=1}^{\sharp \neg ab} c_b^i$ it follows that a knows that b holds at least one more card, namely one of the (other!) cards $c^1, ..., c^m$. Therefore, a can imagine that b holds more than $\sharp b$ cards. From deals (that is known by a) follows that b holds exactly $\sharp b$ cards. Contradiction. □

4.4 Derived characteristics: seedontknow

Using the validity of $K_a \delta_d^a \to \delta_d^a$ for $S5$ models, we can combine seedeal and dontknow in one formula seedontknow. It expresses that players can only imagine deals to be the case that correspond to what they know of their own cards.

Definition 10 (Combining private knowledge and ignorance)

$$\text{seedontknow} := \bigwedge_{a \in A} \bigwedge_{d' \in D_{\sharp d}} (K_a \delta_{d'}^a \leftrightarrow M_a \delta_{d'})$$

5 Description of the pre-initial state

We have described the initial state of a knowledge game, where the cards have been dealt and where players have looked at their cards. Now imagine that the cards have been dealt but that the players have not looked at their cards yet. Now players only know the *size* ($\sharp d$) of the deal: how many cards every player has. Therefore they consider every deal of that size a possibility. This is the *pre-initial state*.

Definition 11 (Pre-initial state) *The* pre-initial state *for a given deal* $d \in A^{\mathbf{C}}$ *is the state* $(preI_d, d)$, *where* $preI_d = \langle D_{\sharp d}, \{U_a\}_{a \in A}, V \rangle$ *with for each player* U_a *the universal relation* $D_{\sharp d} \times D_{\sharp d}$.

Definition 12 (Theory preKgames) *Let $d \in A^{\mathbf{C}}$ be a deal of cards. The theory* preKgames *(for deal d) consists of the constituents:*

$$\begin{aligned}\text{deals} &:= \bigvee\nolimits_{d' \in D_{\sharp d}} \delta_{d'} \\ \text{dontknowany} &:= \bigwedge\nolimits_{a \in A} \bigwedge\nolimits_{d' \in D_{\sharp d}} \neg K_a \neg \delta_{d'}\end{aligned}$$

Theorem 9 *Theory* preKgames *for deal $d \in A^{\mathbf{C}}$ describes $preI_d$.*

We have to show that $preI_d$ is a model of preKgames, which, again, is obvious, and that *only* $preI_d$ models preKgames:

Lemma 10 *Suppose $M \models$ preKgames. Then $M \underline{\leftrightarrow} preI_d$.*

For a proof, see (van Ditmarsch, 2000). Just as in the corollaries 7 and 8, it follows that theory δ_e + preKgames describes $(preI_d, e)$, so that in particular δ_d + preKgames describes $(preI_d, d)$, and that $preI_d \models \varphi \Leftrightarrow preI_d \models C_A \varphi \Leftrightarrow C_A \text{preKgames} \models \varphi \Leftrightarrow C_A \text{preKgames} \models C_A \varphi$, and that preKgames $+ \delta_e \models \psi \Leftrightarrow preI_d, e \models \psi$.

6 Further observations

Fixed point computations We have described some $S5$ models (states) by way of proving that a suggested description (Hexa, Kgames, preKgames) indeed defines the bisimulation class of these models. These descriptions were 'merely' the outcome of a gradual process of generalizing properties of players' knowledge. They can also be directly computed, by a fixed-point construction that applies to finite modal states in general. See (van Benthem, 1998), relating to (Barwise and Moss, 1996). We have applied their construction to game states in (van Ditmarsch, 2000), and proven the results equivalent to the descriptions presented here.

Other game states We have described only the initial state of a knowledge game, and the pre-initial state. Even though we have not given details on how to play knowledge games (for that see (van Ditmarsch, 2001)) it will be obvious that there are many other, possibly more complex, game states for such card games. Of course, future research should include the description of other game states. For many examples of game states, see (van Ditmarsch, 2000).

On the one hand we may expect *concise* descriptions of other game states to be more complex, because they will contain subgroup common knowledge operators. E.g. in $Hexa$, we have that $C_{12} r_1$ (1 and

2 commonly know that 1 holds the red card) is a postcondition of the action where 1 shows red to 2. On the other hand, such descriptions may be equivalent to publicly known theories with only occurrences of *individual* knowledge operators. E.g., in the given example it will suffice to state that $K_2 r_1$ is one of three alternatives, that are commonly known.

Naturally, we would prefer not to have an ad hoc construction each time we are confronted with a new game state, whether this consists of constructing the unique model of a revised theory, or of devising the theory describing an intended model. We can improve on the situation in two ways, one semantically, the other syntactically.

In (van Ditmarsch, 2000), we present a dynamic epistemic language with dynamic modal operators $[\alpha]$ for actions α. In this action language we have described all knowledge game actions. An action is interpreted as a binary relation $[\![\alpha]\!]$ between states, in other words, as a semantic update. We have proven that action execution preserves bisimilarity of states. This provides us with an indirect – and much shorter – proof of lemma 10 that is needed for theorem 9: part of that indirect proof is to show that I_d results from executing an action α in $preI_d$, namely the action look_A corresponding to everybody looking at their cards (van Ditmarsch, 2000). This method could apply to relate other game states too.

Syntactic relativization of formulas (van Benthem, 2000b) may be preferable. There, a procedure is given for the special case of actions that are public announcements. If that procedure could be extended to game actions in general, we could 'directly' revise theories by a rewriting technique transforming constituents φ of the current description Current into constituents φ^α of the description Next of the next game state, such that expressions of type Current $\to [\alpha]$Next are satisfied. For example, we would get preKgames $\to [\mathsf{look}_A]$Kgames. The factual knowledge of deals remains the same, i.e. deals$^{\mathsf{look}_A}$ = deals, but dontknowany $= \bigwedge_{a \in A} \bigwedge_{d' \in D_{\sharp d}} \neg K_a \neg \delta_{d'}$ will be relativized to

$$\text{seedontknow} = \bigwedge_{a \in A} \bigwedge_{d' \in D_{\sharp d}} (K_a \delta^a_{d'} \leftrightarrow \neg K_a \neg \delta_{d'}) ,$$

i.e. dontknowany$^{\mathsf{look}_A}$ = seedontknow. As the action look_A is not a public announcement the current procedure does not apply. We have not pursued this fascinating topic further.

Interpreted systems In section 4 we already paraphrased δ^a_d as the description of the *local state* of player a given deal d. In that sense, a deal of cards can be said to define a global state, with interesting dependencies between local states. Indeed, knowledge game states can be seen as interpreted systems in terms of (Fagin et al., 1995). This is summarily discussed in (van Ditmarsch, 2000), but also calls for further exploration. In particular, game state *frames* appear to have characteristics which correspond (in the technical sense) to multiagent axioms. Also this may help in describing other game states and models.

7 Conclusion

We have described two different states for card games. The model $Hexa$ for the initial game state of three players each holding a card is described by the theory Hexa. The model I_d of an initial game state for a deal d of $|C|$ cards over $|A|$ players is described by the theory Kgames. Similarly, the model $preI_d$ where the players haven't looked at their cards yet is described by the theory preKgames. Descriptions of other game states may be fruitfully pursued from a viewpoint of knowledge relativization.

References

Barwise, J. and L. Moss. 1996. *Vicious Circles*. CSLI Publications, Stanford.

van Benthem, J. 1998. Dynamic odds and ends. Technical report, ILLC, University of Amsterdam. Report ML-1998-08.

van Benthem, J. 2000a. Hintikka self-applied. Manuscript.

van Benthem, J. 2000b. Information update as relativization. Manuscript.

Bonanno, G. 1993. The logical representation of extensive games. *International Journal of Game Theory*, 22(2):153–169.

van Ditmarsch, H. 2000. *Knowledge games*. PhD thesis, University of Groningen. ILLC Dissertation Series DS-2000-06.

van Ditmarsch, H. 2001. Knowledge games. *Bulletin of Economic Research*, 53(4):249–273.

Ehrenfeucht, A. 1961. An application of games to the completeness problem for formalized theories. *Fundamenta Mathematicae*, 49:129–141.

Fagin, R., J. Halpern, Y. Moses, and M. Vardi. 1995. *Reasoning about Knowledge*. MIT Press, Cambridge MA.

Hintikka, J. and G. Sandu. 1989. Informational independence as a semantical phenomenon. In Fenstadt, J., I. Frolov, and R. Hilpinen, editors, *Logic, Methodology and Philosophy of Science VIII*, pages 571–589, Amsterdam. Elsevier.

Kaneko, M. and T. Nagashima. 1996. Game logic and its applications I. *Studia Logica*, 57:325–354.

Meyer, J.-J. and W. van der Hoek. 1995. *Epistemic Logic for AI and Computer Science*. Cambridge Tracts in Theoretical Computer Science 41. Cambridge University Press, Cambridge MA.

Part II
Classical Logic

Resource Consciousness in Classical Logic

Andreas Blass

> Using Herbrand's Theorem, we define simple Herbrand validity, a sort of resource consciousness that makes sense in classical predicate logic. We characterize the propositional formulas all of whose first-order instances are simply Herbrand valid. The characterization turns out to coincide with a known characterization of game semantical validity for multiplicative formulas.

1 Introduction

Resource consciousness is one of the main features of linear and affine logic (Girard, 1987). In these logics, validity of a sequent $\Gamma \vdash A$ means that the conclusion A can be obtained from the list Γ of hypotheses using each hypothesis exactly once (in the case of linear logic) or at most once (in the case of affine logic). Thus, in these logics, a hypothesis that is used twice must occur twice in the list Γ. Classical logic is not resource conscious in this sense, because the tautology $A \to A \wedge A$ allows us to produce as many copies as we want from one hypothesis A.

Nevertheless, we shall show in this paper that a form of resource consciousness can be found in classical first-order logic. Using it, we shall define two strong notions of validity, one for first-order formulas and one for propositional formulas. The new notion of validity in first-order logic is not very well behaved and plays only an auxiliary role, leading up to the propositional notion. The propositional notion, on the other hand, is well behaved, admits a syntactic characterization, and coincides with game semantical validity, as defined in (Blass, 1992), for the multiplicative fragment of affine logic. Except for this

---

[*]Address: Mathematics Department, University of Michigan, Ann Arbor, MI 48109-1109, U.S.A., ablass@umich.edu.
Partially supported by a grant from Microsoft Research

Games, Logic, and Constructive Sets
Grigori Mints and Reinhard Muskens (eds.)
Copyright ©2003, CSLI Publications

introduction and a concluding section, we shall not need any concepts from linear logic or game semantics; the body of this paper is entirely about classical logic.

2 An Example

Consider the sentence τ expressing the transitivity of a binary relation $<$,
$$(\forall x, y, z)\, (x < y \wedge y < z \longrightarrow x < z),$$
and the sentence λ expressing a longer version of transitivity,
$$(\forall x, y, z, w)\, (x < y \wedge y < z \wedge z < w \longrightarrow x < w).$$
Of course, τ logically implies λ. It is intuitively clear that any proof of λ from τ will have to use the hypothesis τ twice, because any particular instance of λ (specific values of x, y, z, w) depends on two instances of τ (either the x, y, z and x, z, w instances or else the y, z, w and x, y, w instances). We shall make this intuition precise and use it to introduce a resource conscious version of logical consequence in which each hypothesis is to be used at most once.

The discussion of τ and λ made essential use of the fact that both of them are universal sentences, so that there is a clear notion of "instance" and we can count how many instances of the hypotheses are needed to establish one instance of the conclusion. One can, of course, give a "dual" discussion in the case of existential sentences. Here the issue will not be the number of uses of the hypothesis but the number of "guesses" of witnesses for the conclusion. Consider, for example, the contrapositive of $\tau \longrightarrow \lambda$, with the negations moved into the scopes of the quantifiers. From
$$(\exists x, y, z, w)\, (x < y \wedge y < z \wedge z < w \wedge \neg x < w)$$
follows
$$(\exists x, y, z)\, (x < y \wedge y < z \wedge \neg x < z).$$
Given a witness x, y, z, w for the hypothesis, we know that either x, y, z or x, z, w serves as a witness for the conclusion, but we do not know which. Our resource conscious version of logical consequence will prohibit this sort of "two attempts at the conclusion" as well as its dual, "two uses of the hypothesis."

The ideas described in this transitivity example can be easily carried over to similar situations involving universal sentences and existential sentences, but it is not immediately clear how to extend them to sentences involving quantifier alternations. The tool that makes such an extension possible is Herbrand's theorem, which we review, in a convenient form, in the next section.

3 Herbrand's Theorem

Herbrand's theorem associates to each first-order sentence ϕ a quantifier-free formula ϕ_H such that ϕ is valid if and only some disjunction of instances of ϕ_H is a tautology.

Convention 3.1 We work in first-order logic without equality. The propositional connectives are $\wedge$, $\vee$ and $\neg$. Negation is applied only to atomic formulas; if we write $\neg$ in any other context, it is to be understood as an "abbreviation" for the result of pushing the negation in to the atomic level, using De Morgan's laws and the usual interchange rules for quantifiers and negations. If we write $A \longrightarrow B$, it is to be understood as an abbreviation of $\neg A \vee B$. We also assume that every vocabulary under consideration has at least one constant symbol.

Remark 3.2 It would make no difference if we used the other common definition of $A \longrightarrow B$ as $\neg(A \wedge \neg B)$, for pushing the negation in to the atomic level would convert this to $\neg A \vee B$. On the other hand, the two common ways of eliminating the biconditional $A \longleftrightarrow B$, namely $(A \longrightarrow B) \wedge (B \longrightarrow A)$ and $(A \wedge B) \vee (\neg A \wedge \neg B)$ are genuinely different. Since we shall be dealing with notions of validity more restrictive than ordinary logical validity, it is not correct to identify logically equivalent formulas, as is often done. See Remark 5.3 below for more about the difference between the two versions of the biconditional. Except for that remark, we shall not use biconditionals again in this paper, so we need not address the question of which interpretation is to be preferred. We thank the referee for pointing out the need for caution on this point.

Definition 3.3 The *Herbrand form* ϕ_H of a first-order sentence ϕ is obtained by the following five steps.

1. Rename bound variables in ϕ so that no variable is quantified twice.

2. Pull out all the universal quantifiers into a prefix of second-order quantifiers, using the usual prenex operations plus the following

equivalences to pull universal quantifiers past existential ones.

$$\exists y\, \forall x\, \alpha(x,y) \longleftrightarrow \forall X\, \exists y\, \alpha(X(y),y)$$
$$\exists y\, \forall X\, \alpha(X(\vec{z}),y) \longleftrightarrow \forall X\, \exists y\, \alpha(X(y,\vec{z}),y).$$

3. Pull out all the existential quantifiers from the first-order part of the formula, using the usual prenex operations. At this point the formula consists of a block of universal second-order quantifiers (over functions), then a block of existential first-order quantifiers, and then a quantifier-free matrix.

4. Delete the universal quantifiers. The function symbols that they quantified are to be added to the vocabulary. So the formula has become an existential first-order sentence in the enlarged vocabulary.

5. Delete the existential quantifiers. The variables $\vec{x}$ that they quantified are the free variables of the Herbrand form ϕ_H, so we often write $\phi_H(\vec{x})$.

Notice that, if a universal quantifier in ϕ lies in the scopes of k existential quantifiers, then the corresponding function symbol in ϕ_H is k-ary, and its arguments, wherever it appears in ϕ_H, are the variables of those k existential quantifiers. If $k = 0$, then we have a 0-ary function symbol, i.e., a constant symbol, in ϕ_H.

Example 3.4 If ϕ is

$$\forall x\, \exists y\, \forall z\, P(x,y,z) \longrightarrow \forall x\, \exists y\, \forall z\, Q(x,y,z),$$

then, remembering that implication is treated as an abbreviation and negations are to be pushed in to the atomic level, we find that the universally quantified variables are the y in the antecedent and the x and z in the consequent. Thus, the Herbrand form ϕ_H is (if we use u, v, w as the new variables in the consequent at step 1)

$$P(x, Y(x), z) \longrightarrow Q(U, v, W(v)).$$

In stating Herbrand's Theorem, we use the usual terminology: "Tautology" means valid in propositional logic. The statement of the theorem also involves the notion of a "closed instance" of a formula. Since three other notions of "instance" will play a role later in the paper, we define all four notions here to avoid confusion.

Definition 3.5

- A *closed instance* of a first-order formula ϕ is obtained by substituting closed terms for all the free variables in ϕ.

- A *first-order instance* of a propositional formula ϕ is obtained by replacing the propositional variables in ϕ by first-order sentences.

- A *propositional instance* of a propositional formula ϕ is obtained by replacing the propositional variables in ϕ by propositional formulas.

- A *variable-merging instance* of a propositional formula ϕ is obtained by replacing the propositional variables in ϕ by propositional variables, not necessarily distinct.

If, as a result of the replacing involved in this definition, the negation symbol is applied to a non-atomic formula, then, in accordance with Convention 3.1, it is understood that the negations are to be pushed in until they apply to atomic formulas.

Theorem 3.6 (Herbrand) *A first-order sentence ϕ is valid if and only if some finite disjunction of closed instances of ϕ_H is a tautology.*

Proof The first three steps in the construction of ϕ_H produce formulas logically equivalent to ϕ. (In the case of the rules for pulling universal quantifiers past existential ones, the soundness of the rules may be easier to see in the dual form,

$$\forall y\, \exists x\, \beta(x,y) \longleftrightarrow \exists X\, \forall y\, \beta(X(y),y),$$

a version of the axiom of choice.) At step 4, where we obtain the sentence $\exists \vec{x}\, \phi_H$, the equivalence is lost, since the truth value of this sentence may depend on the interpretations of the function symbols that had previously been universally quantified. Nevertheless, it is clear that step 4 does not affect logical validity; thus ϕ is logically valid if and only if $\exists \vec{x}\, \phi_H$ is.

Like any existential sentence (in a vocabulary with at least one constant symbol), $\exists \vec{x}\, \phi_H$ is valid if and only if it is true in every "term model," i.e., in every structure where every element is the value of a closed term. The reason is that every structure has a substructure that is a term model, and truth of existential sentences is preserved upward from substructures to superstructures.

A term model amounts to an assignment of truth values to all atomic sentences. So $\exists \vec{x}\, \phi_H$ is true in every term model if and only if every such truth assignment makes at least one closed instance of ϕ_H true. Finally, by the compactness theorem for propositional logic, this is equivalent to the existence of finitely many closed instances of ϕ_H such that every truth assignment makes at least one of them true, i.e., such that their disjunction is a tautology. $\square$

Remark 3.7 Like Herbrand (Herbrand, 1930) but unlike some modern presentations such as (Shoenfield, 1967), we defined the Herbrand form of a sentence without first putting it into (first-order) prenex form. Prenex operations would, in general, move some universal quantifiers into the scopes of more existential quantifiers and would thus complicate the Herbrand form. In the proof of our main theorem, it will be important that these complications do not occur.

Unlike both (Herbrand, 1930) and (Shoenfield, 1967) we have presented Herbrand's theorem as a semantical result, characterizing validity. Herbrand did not work with semantics, and he presented his theorem as a characterization of provability in a certain deductive system. Of course the completeness theorem implies that the two points of view are equivalent, but Herbrand's proof of his theorem was nearly simultaneous with Gödel's proof of the completeness theorem.

Example 3.8 In the previous example,
$$\forall x\, \exists y\, \forall z\, P(x,y,z) \longrightarrow \forall x\, \exists y\, \forall z\, Q(x,y,z)$$
is not logically valid, but it becomes so if we write a second P in place of Q. No disjunction of closed instances of the Herbrand form $P(x, Y(x), z) \longrightarrow Q(U, v, W(v))$ is a tautology (as one can make all atomic formulas with P true and all those with Q false). But if we replace Q with P, the new Herbrand form $P(x, Y(x), z) \longrightarrow P(U, v, W(v))$ has a single instance that is a tautology; replace the variables x, z, and v by the closed terms U, $W(Y(U))$, and $Y(U)$, respectively.

Example 3.9 In the transitivity example from Section 2, the Herbrand form of $\tau \longrightarrow \lambda$ is
$$(x < y \wedge y < z \longrightarrow x < z) \longrightarrow (T < U \wedge U < V \wedge V < W \longrightarrow T < W).$$
Here x, y, z are variables and T, U, V, W are constant symbols. No single closed instance of this formula is a tautology, but there are two closed instances whose disjunction is a tautology. For example, replace x, y, z by T, U, V for the first instance and by T, V, W for the second.

(Alternatively, replace x, y, z by U, V, W for the first instance and by T, U, W for the second.)

Similarly, for the (rephrased) contrapositive

$$(\exists x, y, z, w)\,(x < y \land y < z \land z < w \land \neg x < w) \longrightarrow \\ (\exists x, y, z)\,(x < y \land y < z \land \neg x < z),$$

the Herbrand form is

$$(T < U \land U < V \land V < W \land \neg T < W) \longrightarrow (x < y \land y < z \land \neg x < z),$$

and we obtain a tautologous disjunction of two closed instances by the same substitutions as before.

Notice that the closed instances used in this example correspond exactly to the two uses of the hypothesis and the two attempts at the conclusion in the proofs of $\tau \longrightarrow \lambda$ and its contrapositive, as discussed in Section 2.

4 Simple Herbrand Validity

The preceding example suggests measuring resource usage, or at least bounding it from below, by the number of closed instances needed to produce a tautologous Herbrand disjunction. That is, if, as in the example, two disjuncts are needed, then we regard this as indicating that a hypothesis was used twice or that two attempts were made to obtain witnesses for a conclusion. The following definition is, therefore, intended to capture the idea that no such duplication is needed.

Definition 4.1 A first-order sentence ϕ is *simply Herbrand valid* if some closed instance of its Herbrand form ϕ_H is a tautology.

Thus, for example, $\forall x\, \exists y\, \forall z\, P(x, y, z) \longrightarrow \forall x\, \exists y\, \forall z\, P(x, y, z)$ is simply Herbrand valid, but $\tau \longrightarrow \lambda$ is not (where τ and λ are as in Section 2). The reader is invited to check that $\tau \land \tau \longrightarrow \lambda$ is simply Herbrand valid. Repeating the hypothesis τ makes available, in a single instance of the Herbrand form, the two instances of τ that we need.

Although simple Herbrand validity clearly has something to do with using hypotheses only once (and, dually, making only one guess at a witness for a conclusion), it depends too heavily on the quantifier structure to serve as a really good model of resource consciousness. It is conscious of (and prohibits) the sort of reuse of hypotheses that manifests itself in the occurrence of different instances, but it ignores other sorts of reuse. Thus, for example, $\forall x\,(P(x) \longrightarrow P(x) \land P(x))$ is simply Herbrand valid yet expresses the antithesis of resource consciousness. (A simpler example, if there are no objections to 0-ary predicate symbols, is $P \longrightarrow P \land P$.)

The situation improves if we abstract from the quantifier structure in the following way. Instead of considering a particular first-order formula, consider all first-order instances of some propositional formula.

Definition 4.2 A propositional formula is *universally simply Herbrand valid*, abbreviated *usHv*, if all its first-order instances are simply Herbrand valid.

The following example indicates that this notion captures the idea of resource consciousness better than simple Herbrand validity does. Contrast it with the observation above that $A \longrightarrow A \land A$ is simply Herbrand valid.

Example 4.3 $A \longrightarrow A \land A$ is not usHv. Indeed, its first-order instance

$$\forall x\, \exists y\, P(x,y) \longrightarrow (\forall x\, \exists y\, P(x,y) \land \forall x\, \exists y\, P(x,y))$$

is not simply Herbrand valid. Its Herbrand form is

$$P(x, Y(x)) \longrightarrow P(U, v) \land P(W, z).$$

No single closed instance of this is a tautology. Indeed, in any closed instance, the antecedent matches at most one conjunct from the consequent, so there is a truth assignment verifying the antecedent and falsifying a different conjunct in the consequent. (There are two closed instances whose disjunction is a tautology.)

5 Universal Simple Herbrand Validity

The purpose of this section is to prove the main result of the paper, a syntactic characterization of universal simple Herbrand validity.

Definition 5.1 A propositional formula is *binary* if no propositional variable occurs in it more than twice.

Theorem 5.2 *For any propositional formula ϕ, the following are equivalent.*

1. *ϕ is universally simply Herbrand valid.*

2. ϕ is a propositional instance of a binary tautology.

3. ϕ is a variable-merging instance of a binary tautology.

Proof Obviously, 3 implies 2; the converse is also easy to check directly, but we won't need it because we'll prove 1 $\longrightarrow$ 3 and 2 $\longrightarrow$ 1.
Proof of 1 $\longrightarrow$ 3: Assume ϕ is usHv, and consider the first-order instance ψ obtained by replacing each propositional variable p by a formula of the form $\forall x \, \exists y \, P(x,y)$, using different binary predicate symbols P for different propositional variables p. By assumption, ψ is simply Herbrand valid, so let θ be a tautologous closed instance of ψ_H. Thus, θ is like ϕ except that each occurrence of a propositional variable p has become an occurrence of $P(t,u)$ for some closed terms t and u — possibly different closed terms for different occurrences of the same p.

In fact, we rather rarely get the same $P(t,u)$ repeated. To see this (and to make "rather rarely" precise), recall the first step in the definition of the Herbrand form ψ_H: Rename bound variables so that no variable is quantified twice. Thus, if we consider two positive occurrences of p in ϕ, the corresponding subformulas of ψ will, after this renaming, look like $\forall x \, \exists y \, P(x,y)$ and $\forall x' \, \exists y' \, P(x',y')$, and in the Herbrand form ψ_H these will have become $P(X,y)$ and $P(X',y')$, with different constant symbols X and X' in the first argument place. So these two positive occurrences of p in ϕ will become different atomic formulas in θ.

Similarly, if we consider two negative occurrences of a propositional variable p in ϕ, they become $P(x, Y(x))$ and $P(x', Y'(x'))$ in ψ_H, with different function symbols Y and Y'. So the corresponding atomic formulas in θ are different.

Thus, the only way two atomic formulas in θ can be the same is for one to arise from a positive occurrence and the other from a negative occurrence of some p in ϕ. In particular, no atomic formula can occur three times in θ. If we regard the atomic formulas occurring in θ as propositional variables, then the preceding discussion shows that θ is a binary tautology. Since ϕ is clearly a variable-merging instance of it, obtained by substituting the original variables p for all atomic formulas in θ that begin with the corresponding predicate symbols P, we have established 3.
Proof of 2 $\longrightarrow$ 1: We must show that any first-order instance of any propositional instance of a binary tautology is simply Herbrand valid. The proof proceeds by three reductions of the problem, after which the remaining work is quite easy. The first reduction is to observe that, since "instances of instances are instances" we need only show that

any first-order instance ψ of any binary tautology θ is simply Herbrand valid.

The second reduction is to arrange that, without loss of generality, whenever a propositional variable occurs twice in θ, one occurrence is positive and the other negative. Indeed, suppose p had two positive occurrences, and let θ' be obtained from θ by replacing one of these occurrences by a new propositional variable p'. Clearly, ψ is a first-order instance of θ' and θ' is binary. Furthermore, θ' is a tautology. To see this, suppose we had a truth assignment falsifying θ'. It must give p and p' different truth values, for otherwise it would also falsify θ. But then if we change the value "true" of p or of p' to "false," the new truth assignment will still falsify θ' because both p and p' occurred positively. So the new truth assignment would falsify the tautology θ. This contradiction shows that we can eliminate a repetition of a variable when both its occurrences are positive. An analogous argument eliminates repetitions when both occurrences are negative.

Before proceeding to the third reduction, we summarize the present situation. We assume from now on that θ is a tautology in which each propositional variable has at most one positive occurrence and at most one negative occurrence. Let ψ be any first-order instance of θ, and let ψ_H be its Herbrand form. We shall complete the proof by finding a closed instance of ψ_H that is, when regarded as a propositional formula (viewing its atomic subformulas as propositional variables) a propositional instance of θ and therefore a tautology.

Each propositional variable p in θ becomes some (first-order) subformula π of ψ, which in turn becomes (at worst) two subformulas π^+ and π^- in ψ_H. Here π^+ occurs in place of the positive occurrence of p and π^- occurs in place of the negative occurrence of p in θ. (If p had only one occurrence in θ, then one of $\pi^\pm$ is absent.) Notice that, unless π happened to be quantifier-free, π^+ and π^- are different formulas, for any quantifier in π will be universal in one of the two copies of π and existential in the other. (Remember that negations always get pushed in past the quantifiers.) Our task is to form a closed instance of ψ_H in such a way that the corresponding instances of π^+ and π^- are identical. Then this instance of ψ_H will be an instance of θ (obtained by replacing each propositional variable p in θ by the common instance of the corresponding formulas $\pi^\pm$), so it will be a tautology, and the proof will be complete.

The third reduction is to see that, when defining this instance of ψ_H, we may focus on a single p and the two formulas $\pi^\pm$ that it developed into in going from θ to ψ_H. (If p had only one occurrence in θ, then we need not concern ourselves with it; our only task is to make sure that,

RESOURCE CONSCIOUSNESS IN CLASSICAL LOGIC / 71

when π^+ and π^- both exist, then our closed instantiation makes them identical.) While focusing on one p, we shall define the closed terms that are to be substituted for the free variables in π^+ and π^-; we define them in such a way as to make the resulting instances of π^+ and π^- agree. What we do here for a particular p will not interfere with the corresponding efforts on behalf of other propositional variables q in θ, because those efforts will define what is to be substituted for *different* variables — different because of the renaming step in the definition of Herbrand form. This completes the three reductions.

Concentrating therefore on a particular p, assume that it occurs twice in θ, once positively and once negatively, and assume that its replacement π in ψ has no variable bound twice. (It would be rewritten this way in forming ψ_H, so there is no harm in supposing it is already written this way.) Let $x_1, x_2, \ldots, x_n$ be the variables occurring in the sentence π, ordered in such a way that, if the quantifier of x_i lies in the scope of the quantifier of x_j then $j < i$. For example, the variables could be ordered according to the left-to-right order of occurrence of their quantifiers in π. Notice that, since p occurs twice in θ and π correspondingly occurs twice in ψ, each x_i will become two distinct variables in the construction of ψ_H (see step 1 in the definition of Herbrand form). One of these will be universally quantified, the other existentially. The former becomes, in ψ_H, a function symbol which we call X_i; the latter remains a variable, which we call x'_i. The argument places of X_i are occupied by some x'_j's with smaller subscripts. We write $X_i(x'_1, x'_2, \ldots, x'_{i-1})$, but with the understanding that not all of the indicated arguments have to be present.

Now we define, by induction on i, the closed term t_i to be substituted for the variable x'_i in forming the desired instance of ψ_H. For any particular i, assume by induction that t_j is already defined for all $j < i$. Define t_i to be $X_i(t_1, t_2, \ldots, t_{i-1})$, with the notation as described at the end of the last paragraph (so some of the earlier t_j's may not really be present). It is easy to verify that this definition makes the resulting instances of π^+ and π^- identical. So the proof is complete. □

Remark 5.3 We are now in a position to be more explicit about the difference, already mentioned in Remark 3.2, between the two common definitions of the biconditional,

- $(A \longrightarrow B) \wedge (B \longrightarrow A)$, i.e., $(\neg A \vee B) \wedge (\neg B \vee A)$, and
- $(A \wedge B) \vee (\neg A \wedge \neg B)$.

With the first definition, $A \longleftrightarrow A$ is usHv, because it is an instance of the binary tautology $(\neg A \vee A) \wedge (\neg B \vee B)$. With the second definition,

$A \longleftrightarrow A$ is $(A \wedge A) \vee (\neg A \wedge \neg A)$, which is easily seen not to be an instance of a binary tautology. Thus, the difference between the two definitions of the biconditional is essential in the context of (universal) simple Herbrand validity.

6 Modus Ponens

In the introduction, we mentioned that simple Herbrand validity is not well-behaved and therefore serves mainly as a step on the way to the concept of universal simple Herbrand validity. In this section, we indicate the sort of misbehavior that we had in mind. We give an example of two simply Herbrand valid sentences, of the forms ϕ and $\phi \longrightarrow \psi$, such that ψ is not simply Herbrand valid. That is, the rule of modus ponens is not sound for simple Herbrand validity. We shall see later that it is sound for universal simple Herbrand validity.

We work in a vocabulary with one unary predicate symbol M and, in accordance with Convention 3.1, one constant symbol c. Let ϕ be the sentence
$$\exists u\, M(u) \longrightarrow \exists v\, M(v),$$
and let ψ be
$$\exists x\, \forall y\, (M(y) \longrightarrow M(x)).$$
Then the Herbrand form of ϕ is $M(U) \longrightarrow M(v)$, which has a tautologous closed instance $M(U) \longrightarrow M(U)$.

The Herbrand form of $\phi \longrightarrow \psi$ is
$$(M(u) \longrightarrow M(V)) \longrightarrow (M(Y(x)) \longrightarrow M(x)).$$

This also has a tautologous closed instance, obtained by replacing x with V and u with $Y(V)$.

The Herbrand form of ψ is $M(Y(x)) \longrightarrow M(x)$. In any closed instance of this, the antecedent is different from the consequent, so no single closed instance of ψ_H is a tautology.

Thus, both ϕ and $\phi \longrightarrow \psi$ are simply Herbrand valid but ψ is not.

In this example, although no single closed instance of ψ_H is a tautology, there are two instances whose disjunction is a tautology. Replace x by c for one instance and by $Y(c)$ for the other.

The example can be modified so that ψ is much farther from being simply Herbrand valid. Given any positive integer s, work in a vocabulary having s unary predicate symbols $M_1, \ldots, M_s$ and one constant symbol, let ϕ be

$$(\exists u_1\, M_1(u_1) \longrightarrow \exists v_1\, M_1(v_1)) \wedge \cdots \wedge (\exists u_s\, M_s(u_s) \longrightarrow \exists v_s\, M_s(v_s)),$$

and let ψ be

$$\exists x_1 \ldots \exists x_s \, \forall y \, [\neg M_1(y) \vee (M_1(x_1) \wedge \neg M_2(y)) \vee \ldots$$
$$\ldots \vee (M_{s-1}(x_{s-1}) \wedge \neg M_s(y)) \vee M_s(x_s)].$$

Then both ϕ and $\phi \longrightarrow \psi$ are simply Herbrand valid, but a tautologous disjunction of closed instances of ψ_H must use at least $s+1$ instances.

7 Connection with Affine Logic

In (Blass, 1992), I proposed a game semantics for Girard's affine logic (Girard, 1987) and proved a soundness theorem for it. For the additive fragment, there is a completeness theorem also (Blass, 1992, Section 4), but this is not so for the multiplicative fragment. The game-valid multiplicative formulas are a strictly wider class than the ones provable in affine logic. The former class admits, however, a syntactic characterization, given by the "Multiplicative Validity Theorem" of (Blass, 1992, Section 5): Up to notational differences (the $A^\perp$ and $A \otimes B$ of affine logic corresponding to the $\neg A$ and $A \wedge B$ of traditional logic), the game-valid multiplicative formulas are exactly the propositional instances of binary tautologies. That is, they are exactly the usHv formulas.

There is some empirical evidence for the naturalness of the game semantics of (Blass, 1992). First, the operations on games used to interpret the connectives arose originally in purely game-theoretic considerations (Blass, 1972). Second, they arose independently in Japaridze's analysis (Japaridze, 1997) of "effective truth." The games and strategies considered in (Japaridze, 1997) are quite different from those in (Blass, 1992), for in Japaridze's games all plays are finite and strategies are required to be effectively computable. Nevertheless, Japaridze finds the same notion of validity for the multiplicative fragment.

That this same notion of validity has now arisen a third time, from the analysis of Herbrand disjunctions, increases the evidence for the importance of these concepts. It suggests that instances of binary tautologies occur naturally in attempts to analyze resource consciousness and deserve more attention in their own right. I am aware of only one paper, Jaśkowski's (Jaśkowski, 1963), in which binary tautologies are studied for their own sake. They occur there as the solution to the problem of characterizing the provable formulas of a certain deductive system.

It is well known that the new function symbols occurring in the Herbrand form ϕ_H of a sentence ϕ can be interpreted as (part of) a strategy for player $\forall$ in a game where $\exists$ tries to confirm the truth of ϕ

while ∀ tries to refute ϕ. This might lead one to regard a connection between Herbrand's theorem and game semantics as unsurprising. I know of no way, however, to make these ideas precise. In particular, I do not know how to use the relatively easy proof of Theorem 5.2 in the present paper to simplify the considerably harder proof of the multiplicative validity theorem in (Blass, 1992).

Finally, it should be pointed out that, in contrast to what we showed in Section 6 for simple Herbrand validity, universal simple Herbrand validity is preserved by modus ponens and by the slightly more general cut rule. A proof using too much machinery is to apply the soundness theorem of (Blass, 1992) which says that cut (along with all the other rules of affine logic) is sound for game-validity, and to invoke the fact that game-validity in the multiplicative fragment coincides with universal simple Herbrand validity. An alternate proof is to show by a direct combinatorial argument that cut preserves the property of being an instance of a binary tautology; that combinatorial argument is left as an amusing exercise for the reader (as it was in (Blass, 1992)).

References

Blass, A. 1972. Degrees of indeterminacy of games. *Fundamenta Mathematicae*, 77:151–166.

Blass, A. 1992. A game semantics for linear logic. *Annals of Pure and Applied Logic*, 56:183–220.

Girard, J.-Y. 1987. Linear logic. *Theoretical Computer Science*, 50:1–102.

Herbrand, J. 1930. *Recherches sur la théorie de la démonstration.* Number 33 in Travaux de la Société des Sciences et Lettres de Varsovie, Classe III, sciences mathématiques et physiques.

Japaridze, G. 1997. A constructive game semantics for the language of linear logic. *Annals of Pure and Applied Logic*, 85:87–156.

Jaśkowski, S. 1963. Über Tautologien, in welchen keine Variable mehr als zweimal vorkommt. *Zeitschrift für mathematische Logik und Grundlagen der Mathematik*, 9:219–228.

Shoenfield, J. R. 1967. *Mathematical Logic*. Addison-Wesley.

Quick Cut-Elimination for Monotone Cuts

Grigori Mints

> Using pruning transformations of derivations, we present a short proof that cut on *monotone* formulas (that do not contain implication or negation) can be eliminated much quicker and with much smaller blow-up than in a general case. This is a significant simplification of the proof and algorithm given earlier by M. Baaz and A. Leitsch.

1 Introduction

It is well-known that reduction of logical proofs to a normal form can be interpreted as a computation of a value of a program, and this interpretation leads to many fruitful ideas and results. For L-style systems, that have rules for introduction of logical connectives both into *antecedent* (to the left of an arrow) and into *succedent* (to the right of the arrow) normalization means cut-elimination. In general cut-elimination is hyper-exponential in the worst case: for a given proof d the normal form of d can be of the height $2_c(h)$ where

$$2_1(n) = 2^n, 2_{k+1}(n) = 2^{2_k(n)}$$

h is the height of d, and c is the maximum complexity of cut formulas. In this note we present a short proof that cut on *monotone* formulas (that do not contain $\to, \neg$) can be eliminated much quicker and with much smaller blow-up. This is a significant simplification of the proof and algorithm in Baaz and Leitsch (1999) where it was first established that all monotone cuts can be quickly (almost exponentially) eliminated from proofs (what is called there) quasi-monotone sequents. Namely, for such a proof d there is a cut-free proof d' of the same sequent with

$$l(d') < size(d) \cdot l(d) \cdot 2^{l(d)} \tag{1}$$

---

*Address: Department of Philosophy, Stanford University, Stanford, CA 94305, U.S.A., mints@turing.stanford.edu.

Games, Logic, and Constructive Sets
Grigori Mints and Reinhard Muskens (eds.)
Copyright ©2003, CSLI Publications

where $size(d)$ is the maximum complexity of formulas in d, and $l(d)$ is the number of sequents in d. Our proof provides a bound

$$l(d') \leq l(d) \cdot ax_1 \cdot \ldots \cdot ax_k \qquad (2)$$

where ax_i is the maximum of 1 and the number of axioms traceable to the antecedent formula of the i-th cut. Since

$$ax_1 \cdot \ldots \cdot ax_k < 2^{ax_1} \cdot \ldots \cdot 2^{ax_k} = 2^{ax_1+\ldots+ax_k}$$

and $ax_1 + \ldots + ax_k < l(d)$, bound (2) implies (1). The latter bound is exponential in $l(d)$. Although this is optimal in the worst case as an example in Baaz and Leitsch (1999) shows, (2) is linear in $l(d)$ with a multiplier depending of the part to be deleted in the cut-elimination process.

Our cut-elimination algorithm works for a smaller class: proofs that have cuts on monotone formulas and endsequents

$$H \Rightarrow M, \qquad (3)$$

where H is a finite sequence of Horn formulas, and M is a monotone formula. Let us recall that a *universal Horn formula* is a formula of the form

$$\forall \mathbf{x}(P_1 \to \ldots \to (P_n \to P)\ldots),$$

where P_i, P are atomic formulas, $\mathbf{x} = x_1, \ldots, x_n$ is a finite sequence of variables.

Our restriction still allows to prove (exactly as in Baaz and Leitsch (1999)) that monotone formulas do not define a cut-reduction class. This means that there is no polynomial reduction of arbitrary proofs to proofs containing only monotone cuts. Indeed, Statman's sequence of equational sequents with no Kalmar-elementary bound for cut-elimination Statman (1979) can be encoded by sequents of the form (3) since the standard properties of equality can be encoded by Horn formulas, after which equality can be replaced by a new binary predicate.

Our simplified cut-elimination uses pruning transformations (similar to ones underlying Harrop's theorem) employed in Mints (1974). They allow drastically "skolemize" formulas that are proved from Horn axioms. Let a *majorant* of a monotone formula A be any formula obtained from A by retaining exactly one of the disjuncts in any disjunction and instantiating existential quantifiers by arbitrary terms and universal quantifiers by fresh variables (which do not occur in A). It is easy to see that for every majorant M' of a formula M one can prove

$$\forall \mathbf{x} M' \to M$$

where **x** is a sequence of variables introduced to majorize universal quantifiers. Our cut-elimination procedure is based on the following statement which is easily derived from Lemma 3.1 below.

Let H be a finite sequence of universal Horn formulas, **M** be a monotonic formula. Then every cut-free proof of $H \Rightarrow M$ can be pruned (by deleting formulas, sequents and applications of the rules) into a cut-free proof of $H \Rightarrow M'$ for some majorant M' of the formula M.

This allows reduction to replace an exponential blow-up in one step of the standard cut-elimination method by a multiplicative increase, see Lemma 3.2. Iteration of this reduction produces the bound (2).

Two principal differences with the proof in Baaz and Leitsch (1999) concern the treatment of existence and disjunction. Our restriction to Horn antecedents in endsequents (3) allows us to avoid skolemization and subsequent reskolemization. On the other hand, we apply "inversion" of the left-hand side premise of $\vee$-cut only locally, that is when a cut-formula is of the form $A \vee B$. In Baaz and Leitsch (1999) disjunction is eliminated from the inside of cut-formulas "globally". As a referee pointed out, the confluence status of this method is not clear. However, when optimization is a goal, other criteria are often more important. Moreover, as pointed out by Kreisel, non-confluence is a bonus, since it allows us to choose a "better" proof among several possible normal forms.

In proofs by induction IH stands for induction hypothesis.

2 System LK

We slightly change the standard definition of structural rules: A series of weakenings or contractions is now counted as one weakening or contraction.

Axioms: $A \Rightarrow A$; $\bot \Rightarrow A$, for atomic A.

Inference Rules

$$\wedge : r \; \frac{\Gamma \Rightarrow \Delta, A \quad \Pi \Rightarrow \Lambda, B}{\Gamma, \Pi \Rightarrow \Delta, \Lambda, A \wedge B} \qquad \vee : l \; \frac{A, \Gamma \Rightarrow \Delta \quad B, \Pi \Rightarrow \Lambda}{A \vee B, \Gamma, \Pi \Rightarrow \Delta, \Lambda}$$

$$\wedge : l1 \; \frac{A, \Gamma \Rightarrow \Delta}{A \wedge B, \Gamma \Rightarrow \Delta} \qquad \wedge : l2 \; \frac{B, \Gamma \Rightarrow \Delta}{A \wedge B, \Gamma \Rightarrow \Delta}$$

$$\vee : r1 \; \frac{\Gamma \Rightarrow \Delta, A}{\Gamma \Rightarrow \Delta, A \vee B} \qquad \vee : r2 \; \frac{\Gamma \Rightarrow \Delta, B}{\Gamma \Rightarrow \Delta, A \vee B}$$

$$\rightarrow : l \; \frac{\Gamma \Rightarrow \Delta, A \quad B, \Pi \Rightarrow \Lambda}{A \rightarrow B, \Gamma, \Pi \Rightarrow \Delta, \Lambda} \qquad \rightarrow : r \; \frac{A, \Gamma \Rightarrow \Delta, B}{\Gamma \Rightarrow \Delta, A \rightarrow B}$$

$$\neg : r \frac{A, \Gamma \Rightarrow \Delta}{\Gamma \Rightarrow \Delta, \neg A} \qquad \rightarrow : l \frac{\Gamma \Rightarrow \Delta, A}{\neg A, \Gamma \Rightarrow \Delta}$$

$$\forall : r \frac{\Gamma \Rightarrow \Delta, A[x/\alpha]}{\Gamma \Rightarrow \Delta, \forall x A} \qquad \forall : l \frac{A[x/t], \Gamma \Rightarrow \Delta}{\forall x A, \Gamma \Rightarrow \Delta}$$

$$\exists : r \frac{\Gamma \Rightarrow \Delta, A[x/t]}{\Gamma \Rightarrow \Delta, \exists x A} \qquad \exists : l \frac{A[x/\alpha], \Gamma \Rightarrow \Delta}{\exists x A, \Gamma \Rightarrow \Delta}$$

The structural rules

$$w : r \frac{\Gamma \Rightarrow \Delta}{\Gamma \Rightarrow \Delta, \Lambda} \qquad w : l \frac{\Gamma \Rightarrow \Delta}{\Lambda, \Gamma \Rightarrow \Delta}$$

$$c : r \frac{\Gamma \Rightarrow \Delta, \Lambda, \Lambda}{\Gamma \Rightarrow \Delta, \Lambda} \qquad c : l \frac{\Lambda, \Lambda, \Gamma \Rightarrow \Delta}{\Lambda, \Gamma \Rightarrow \Delta}$$

$$cut \frac{\Gamma \Rightarrow \Delta, A \quad A, \Pi \Rightarrow \Lambda}{\Gamma, \Pi \Rightarrow \Delta, \Lambda}$$

with standard proviso for *eigenvariable* α and restriction for substitution $[x/t]$.

We consider proofs in classical predicate logic LK. Notation

$$d : \Gamma \Rightarrow \Delta$$

indicates that d is a proof of the sequent $\Gamma \Rightarrow \Delta$. $\Gamma \vdash \Delta$ means derivability of $\Gamma \Rightarrow \Delta$. Length $l(d)$ of a proof d is the number of sequents in d.

3 Cut-Elimination

An *immediate majorant* of a formula M is M itself if it is atomic or of the form $A \wedge B$, each of A_1, A_2 if $M = A_1 \vee A_2$, every formula $A[x/t]$ if $M = \exists x A$, and every formula $A[x/\alpha]$ with $\alpha \notin FV(M)$ if $M = \forall x A$.

Lemma 3.1 (Inversion Lemma) *(a) Let H be a finite sequence of universal Horn formulas, $\mathbf{M}$ be a finite sequence of monotonic formulas. Then every cut-free proof of*

$$H \Rightarrow \mathbf{M} \qquad (4)$$

can be pruned (by deleting formulas, sequents and inferences) into a cut-free intuitionistic proof of $H \Rightarrow M'$ for some immediate majorant M' of M for some $M \in \mathbf{M}$.

(b) Moreover, if $M = \forall x M_1$, then one can choose $M' = M_1[x/\alpha]$ for every $\alpha \notin FV(H, M)$.

(c) If $d : \Gamma \Rightarrow \Delta, A_1 \wedge A_2$ then we can construct $d_i : \Gamma \Rightarrow \Delta, A_i$ ($i = 1, 2$) by pruning d.

Proof. Consider (a),(b) first. We use induction on the given proof of (4) and the fact that $\vee, \exists$-antecedent are not present. By the subformula property and the definition of a Horn formula, all sequents occurring in a cut-free proof of (4) have the same form: Horn antecedent and a sequence of monotone formulas in the succedent. Induction base, when (4) is an axiom $M \Rightarrow M$, is obvious: $M' = M$. For induction step consider the last rule L of the given proof and apply IH to its premises. If the distinguished component M in at least one premise is not the side formula of the rule L, choose the same component M and the same majorant M' for the conclusion, and discard L if it was a succedent rule. If L is a succedent logical rule and the distinguished components in all premises are side formulas, we get an immediate majorant for the principal formula. For example,

$$L = \forall : r \quad \frac{H \Rightarrow \mathbf{M}, M_1[x/\alpha]}{H \Rightarrow \mathbf{M}, \forall x M_1}$$

is pruned into

$$H \Rightarrow M_1[x/\alpha]$$

Renaming, if needed, eigenvariables of other inferences and bound variables, we can choose α to be any variable not free in $H, \forall x M_1$. The only remaining rule is $\rightarrow : l$:

$$\frac{\Gamma \Rightarrow \Delta, A \quad B, \Pi \Rightarrow \Lambda}{A \rightarrow B, \Gamma, \Pi \Rightarrow \Delta, \Lambda}$$

Note that A is an atomic formula since $A \rightarrow B$ is a Horn formula. If the majorant provided by IH for the left premise $\Gamma \Rightarrow \Delta, A$ is not A, take it for the conclusion. Otherwise we can take the majorant from the right premise $B, \Pi \Rightarrow \Lambda$. This proves (a),(b).

(c) is well-known. It is established by a similar induction on given proof. Consider only the case, when the proof ends in antecedent contraction with the principal formula $A_1 \wedge A_2$:

$$c : r \quad \frac{d' : \Gamma \Rightarrow \Delta, \Lambda, A_1 \wedge A_2, \Lambda, A_1 \wedge A_2}{d : \Gamma \Rightarrow \Delta, \Lambda, A_1 \wedge A_2}$$

Apply IH twice to d' to get $d'_i : \Gamma \Rightarrow \Lambda, A_i, \Lambda, A_i$, and apply $c : r$ to get $d_i : \Gamma \Rightarrow \Delta, \Lambda, A_i$. □

We say that a proof d' has *at most the same cut inferences and axioms traceable to cuts* as a proof d if there is (a) an injection of the set of cut-inferences of d' into the set of cut-inferences of d, and (b) an injection of the set of occurrences of axioms in d' traceable to cuts to the set of occurrences of axioms in d traceable to cuts such that corresponding cuts in (a) have the same cut-formulas up to substitution of terms for variables, and similarly for (b). Λ^n stands for n copies of a formula sequence Λ. The next Lemma will be applied with $k = 1, n_1 = 1$, see (5) below.

Lemma 3.2 (Eliminating cuts in parallel) *Let H be a finite sequence of universal Horn formulas, $M_1, \ldots, M_k$ be a finite sequence of monotone formulas,*

$$d_i : H \Rightarrow M_i \ (i = 1, \ldots, k)$$

be cut-free proofs, and

$$e : M_1^{n_1}, \ldots, M_k^{n_k}, \Gamma \Rightarrow \Delta \ (n_1 \geq 0, \ldots, n_k \geq 0).$$

Then there is a proof

$$f : H^{n_1 + \ldots + n_k}, \Gamma \Rightarrow \Delta$$

with at most the same cut inferences as in e and

$$l(f) \leq ax_1 \cdot (l(d_1) - 1) + \ldots + ax_k \cdot (l(d_k) - 1) + l(e)$$

where
$$ax_i := \text{ the number of axioms in } e \text{ traceable to the antecedent } M_i.$$

Proof. Induction on $l(e)$. In fact we prove admissibility of the following series of parallel cuts:

$$\frac{\mathbf{d} : H \Rightarrow \mathbf{M} \quad e : \mathbf{M^n}, \Gamma \Rightarrow \Delta}{f : H^\mathbf{n}, \Gamma \Rightarrow \Delta} \tag{5}$$

and the bound
$$l(f) \leq \mathbf{ax} \cdot (l(\mathbf{d}) - 1) + l(e) \tag{6}$$

where $\mathbf{d} : H \Rightarrow \mathbf{M}$ stands for $d_i : H \Rightarrow M_i \ (i = 1, \ldots, k)$, $e : \mathbf{M^n}, \Gamma \Rightarrow \Delta$ stands for $e : M_1^{n_1}, \ldots, M_k^{n_k}, \Gamma \Rightarrow \Delta$, $f : H^\mathbf{n}, \Gamma \Rightarrow \Delta$ stands for $f : H^{n_1 + \ldots + n_k}, \Gamma \Rightarrow \Delta$.

We move (5) up along e till it disappears at a weakening or axiom. At the places where standard cut-elimination procedures stop after replacing a cut by one or several cuts of smaller complexity, we add such simpler cut to our list and continue.

If e is an axiom $M_1 \Rightarrow M_1$, then $f = d_1, k = 1, ax_1 = 1$ and $l(f) = l(d_1) = 1 \cdot (l(d_1) - 1) + 1$.

In the remaining cases, let L be the last inference in e.

Case 1. The principal formula of L is different from M.

Case 1.1 One-premise rule, say $\vee : r$. Original proof e is shown at the left below, the new proof f is shown at the right. The proof f' is obtained from e' by IH:

$$\vee : r \; \frac{e' : \mathbf{M^n}, \Gamma \Rightarrow \Delta, A}{e : \mathbf{M^n}, \Gamma \Rightarrow \Delta, A \vee B} \qquad \vee : r \; \frac{f' : H^\mathbf{n}, \Gamma \Rightarrow \Delta, A}{f : H^\mathbf{n}, \Gamma \Rightarrow \Delta, A \vee B}$$

Now

$$l(f) = l(f') + 1 \leq \mathbf{ax} \cdot (l(\mathbf{d}) - 1) + l(e') + 1 = \mathbf{ax} \cdot (l(\mathbf{d}) - 1) + l(e) \quad (7)$$

as required.

Case 1.2 Two-premise rule, say $\wedge : r$:

$$\wedge : r \; \frac{e' : \mathbf{M^p}, \Gamma \Rightarrow \Delta, A \quad e'' : \mathbf{M^q}, \Pi \Rightarrow \Lambda, B}{e : \mathbf{M^n}, \Gamma, \Pi \Rightarrow \Delta, \Lambda, A \wedge B}$$

$$\frac{f' : H^\mathbf{p}, \Gamma \Rightarrow \Delta, A \quad f'' : H^\mathbf{q}, \Pi \Rightarrow \Lambda, B}{f : H^\mathbf{n}, \Gamma, \Pi \Rightarrow \Delta, \Lambda, A \wedge B}$$

where $\mathbf{n} = \mathbf{p} + \mathbf{q}$. Let $d' : H \Rightarrow \mathbf{M^p}, d'' : H \Rightarrow \mathbf{M^q}$ and $\mathbf{ax'}(\mathbf{ax''})$ is the number of axioms in e' (in e'') traceable to $\mathbf{M^p}$ (to $\mathbf{M^p}$). Then $\mathbf{ax} = \mathbf{ax'} + \mathbf{ax''}$ and

$$l(f) = l(f') + l(f'') + 1$$
$$\leq \mathbf{ax'} \cdot (l(\mathbf{d'}) - 1) + l(e') + \mathbf{ax''} \cdot (l(\mathbf{d''}) - 1) + l(e'') + 1$$
$$\leq \mathbf{ax} \cdot (l(\mathbf{d}) - 1) + l(e)$$

as required. If, for example, $\mathbf{p} = 0$, then $\mathbf{ax'} = 0$ and the previous computation still goes through.

Case 2. The principal formula of L belongs to $\mathbf{M}$.

Case 2.1. $L = w : l$. Then we have

$$\frac{e' : \mathbf{M'^q}, \Gamma \Rightarrow \Delta}{e : \mathbf{M^n}, \Gamma \Rightarrow \Delta} \qquad \frac{f' : H^\mathbf{q}, \Gamma \Rightarrow \Delta}{e : H^\mathbf{n}, \Gamma \Rightarrow \Delta}$$

with $\mathbf{M'} \subset \mathbf{M}, \mathbf{q} \leq \mathbf{n}$. For $d' : \mathbf{M'}$ which coincides with $\mathbf{d}$ or is obtained by removing some proofs from $\mathbf{d}$, we have

$$l(f) = l(f') + 1 \leq \mathbf{ax'} \cdot (l(\mathbf{d'}) - 1) + l(e') + 1 = \mathbf{ax} \cdot (l(\mathbf{d'}) - 1) + l(e)$$
$$= \mathbf{ax} \cdot (l(\mathbf{d}) - 1) + l(e)$$

as required.

Case 2.2. L is $c : l$:

$$\dfrac{e' : \mathbf{M^{2q}}, \Lambda, \Lambda, \mathbf{M^p}, \Gamma \Rightarrow \Delta}{e : \mathbf{M^q}, \Lambda, \mathbf{M^p}, \Gamma \Rightarrow \Delta} \qquad \dfrac{f' : H^{2q}, \Lambda, \Lambda, H^p, \Gamma \Rightarrow \Delta}{e : H^q, \Lambda, H^p, \Gamma \Rightarrow \Delta}$$

The estimate for $l(f)$ is obtained as in (7).

Case 2.3. $L = \vee : l$.

$$\dfrac{e' : M', \mathbf{M^p}, \Gamma \Rightarrow \Delta \quad e'' : M'', \mathbf{M^q}, \Pi \Rightarrow \Lambda}{e : M' \vee M'', \mathbf{M^{p+q}}, \Gamma, \Pi \Rightarrow \Delta, \Lambda}$$

Assuming that $M' \vee M'' = M_1$ and the immediate majorant in Lemma 3.1 (a) is M', there is $d_1' : H \Rightarrow M'$ with $l(d_1') \leq l(d_1)$. Let $\mathbf{d_2} := d_2 \ldots d_k$ and let f' be obtained by IH from $d_1', \mathbf{d_2}$ and e':

$$w : l, w : r \ \dfrac{f' : H_1, H^p, \Gamma \Rightarrow \Delta}{f : H_1, H^{p+q}, \Gamma, \Pi \Rightarrow \Delta, \Lambda}$$

Note that $l(e) = l(e') + l(e'') + 1 \geq l(e') + 2$. Let $\mathbf{ax_1}(\mathbf{ax_2})$ be the number of axioms in e' traceable to M' (to $\mathbf{M^p}$). We have:

$$l(f) \leq l(f') + 2 \leq \mathbf{ax_1} \cdot (l(d_1) - 1) + \mathbf{ax_2} \cdot (l(\mathbf{d_2}) - 1) + l(e') + 2$$
$$\leq \mathbf{ax} \cdot (l(\mathbf{d}) - 1) + l(e)$$

as required.

Case 2.4. L is one-premise rule, say $\forall : l$.

$$\dfrac{e' : M_1[x/t], \mathbf{M^n}, \Gamma \Rightarrow \Delta}{e : \forall x M_1, \mathbf{M^n}, \Gamma \Rightarrow \Delta}$$

From $d_1 : H \Rightarrow \forall x M_1$ by Lemma 3.1(b) we obtain a proof of $d'' : H \Rightarrow M_1[x/\alpha]$ with a fresh variable α and $l(d'') \leq l(d_1)$. Substitution $[\alpha/t]$ produces $d' : H \Rightarrow M_1[x/t]$ with $l(d') = l(d'') \leq l(d_1)$. The proof $f : H_1, \mathbf{M^n}$ is obtained by IH, and the estimate for $l(e)$ is proved as in (7). □

Theorem 3.1 *Let $g : H \Rightarrow \mathbf{M}$ for a sequence $\mathbf{M}$ of monotonic formulas contains exactly k cuts, all with monotone cut-formulas. Then there is a cut-free $\tilde{g} : H \Rightarrow \mathbf{P}$ with*

$$l(\tilde{g}) \leq l(g) \cdot ax_1 \cdot \ldots \cdot ax_k \tag{8}$$

where ax_i is the maximum of 1 and the number of axioms traceable to the antecedent formula of the i-th cut.

Proof. We eliminate the leftmost cuts using the previous Lemma. More precisely, we use induction on k with subsidiary induction on g.

Induction base $k = 0$ is trivial.

For induction step consider the last inference L of g. Assume first that L is not a cut and has two premises proved by g_1 (with p cuts) and g_2 (with q cuts). We have $k = p + q$ and

$$\begin{aligned} l(\tilde{g}) &= l(\tilde{g}_1) + l(\tilde{g}_2) + 1 \\ &\leq l(g_1) \cdot ax_1 \cdot \ldots \cdot ax_p + l(g_2) \cdot ax_{p+1} \cdot \ldots \cdot ax_{p+q} \\ &< (l(g_1) + l(g_2) + 1) \cdot ax_1 \cdot \ldots \cdot ax_{p+q} \end{aligned}$$

Assume now that L is a cut:

$$\frac{d : H \Rightarrow M \quad e : m, G \Rightarrow \mathbf{M}}{g : H, G \Rightarrow \mathbf{M}}$$

where d contains first p cuts, L is the $(p+1)$-st cut, e contains remaining q cuts, $k = p + q + 1$ and $l(g) = l(d) + l(e) + 1$. By IH we get a cut-free $\tilde{d} : H \Rightarrow M$ with

$$l(\tilde{d}) < l(d) \cdot ax_1 \cdot \ldots \cdot ax_p$$

By Lemma 3.2 we get $f : H, G \Rightarrow \mathbf{M}$ with q cuts and

$$\begin{aligned} l(f) &\leq l(\tilde{d}) \cdot ax_1 \cdot \ldots \cdot ax_p + l(e) \\ &\leq l(d) \cdot ax_1 \cdot \ldots \cdot ax_{p+1} + l(e) \end{aligned}$$

By IH we get cut-free $\tilde{f} : H, G \Rightarrow \mathbf{M}$ with

$$\begin{aligned} l(\tilde{f}) &< l(f) \cdot ax_{p+2} \cdot \ldots \cdot ax_{p+q+1} \\ &\leq l(d) \cdot ax_1 \cdot \ldots \cdot ax_{p+q} \cdot ax_{p+2} \cdot \ldots \cdot ax_{p+q+1} + \\ &\quad l(e) \cdot ax_{p+2} \cdot \ldots \cdot ax_{p+q+1} \\ &< (l(d) + l(e) + 1) \cdot ax_1 \cdot \ldots \cdot ax_{p+q+1} \end{aligned}$$

as required. □

References

Baaz, M. and A. Leitsch. 1999. Cut normal forms and proof complexity. *APAL*, 97:127–177.

Mints, G. 1974. The Skolem Method in Intuitionistic Calculi. In *Proc. Inst. Steklov*, volume 121, pages 73–103. AMS.

Statman, R. 1979. Lower bounds on Herbrand's Theorem. In *Proc. AMS*, volume 75, pages 104–107.

Part III

Constructive Set Theory

The Anti-Foundation Axiom in Constructive Set Theories

Michael Rathjen

> The paper investigates the strength of the anti-foundation axiom on the basis of various systems of constructive set theories.

1 Introduction

Intrinsically circular phenomena have come to the attention of researchers in differing fields such as mathematical logic, computer science, artificial intelligence, linguistics, cognitive science, and philosophy. Logicians first explored set theories whose universe contains what are called non-wellfounded sets, or hypersets (cf. Forti and Honsell (1983), Aczel (1988)). But the area was considered rather exotic until these theories were put to use in developing rigorous accounts of circular notions in computer science (cf. Barwise and Moss (1996)). Instead of the Foundation Axiom these set theories adopt the so-called *Anti-Foundation Axiom,* **AFA**, which gives rise to a rich universe of sets. **AFA** provides an elegant tool for modeling all sorts of circular phenomena. The application areas range from knowledge representation and theoretical economics to the semantics of natural language and programming languages.

The subject of hypersets and their applications is thoroughly and timely developed in the book Barwise and Moss (1996). While reading Barwise and Moss (1996), I asked myself whether most of the material could be developed on the basis of a constructive universe of hypersets rather than a classical one. My tentative answer is that large chunks of

---

*Department of Pure Mathematics, University of Leeds, Leeds LS2 9JT, United Kingdom, E-mail: rathjen@amsta.leeds.ac.uk
Supported by the Volkswagen-Stiftung (RiP program Oberwolfach). Most of the results reported in this paper where first presented in a talk given in the Stockholm-Uppsala Logic Seminar in January 1999.

Games, Logic, and Constructive Sets
Grigori Mints and Reinhard Muskens (eds.)
Copyright ©2003, CSLI Publications

Barwise and Moss (1996) only require constructive set theory. Research in this direction is under way. In the meantime the present paper investigates the strength of **AFA** on the basis of various systems of constructive set theories, including ones with large set axioms.

Constructive set theory grew out of Myhill's endeavours (cf. Myhill (1975)) to discover a simple formalism that relates to Bishop's constructive mathematics as **ZFC** relates to classical Cantorian mathematics. Later on Aczel modified Myhill's set theory to a system which he called Constructive Zermelo-Fraenkel set theory, **CZF**, and corroborated its constructiveness by interpreting it in Martin-Löf type theory (**MLTT**) (cf. Aczel (1978)). The interpretation was in many ways canonical and can be seen as providing **CZF** with a standard model in type theory.

Let **CZF**$^-$ be **CZF** without $\in$-induction and let **CZFA** be **CZF**$^-$ plus **AFA**. I. Lindström (cf. Lindström (1989)) showed that **CZFA** can be interpreted in **MLTT** as well. Among other sources, the work of Lindström (1989) will be utilized in calibrating the exact strength of various extensions of **CZFA**, in particular ones with inaccessible set axioms. The upshot is that **AFA** does not yield any extra proof-theoretic strength on the basis of constructive set theory and is indeed much weaker in proof strength than $\in$-Induction. This contrasts with Kripke-Platek set theory, **KP**. The theory **KPA**, which adopts **AFA** in place of the Foundation Axiom scheme, is proof-theoretically considerably stronger than **KP** as was shown in Rathjen (2001).

This paper also contains several other new results.

2   The anti-foundation axiom

Definition 2.1 A *graph* will consist of a set of *nodes* and a set of *edges*, each edge being an ordered pair (x, y) of nodes. If (x, y) is an edge then we'll write $x \to y$ and say that y is a *child* of x.

A *path* is a finite or infinite sequence $x_0 \to x_1 \to x_2 \to \ldots$ of nodes $x_0, x_1, x_2, \ldots$ linked by edges $(x_0, x_1), (x_1, x_2), \ldots$.

A *pointed graph* is a graph together with a distinguished node x_0 called its *point*. A pointed graph is *accessible* if for every node x there is a path $x_0 \to x_1 \to x_2 \to \ldots \to x$ from the point x_0 to x.

A *decoration* of a graph is an assignment d of a set to each node of the graph in such a way that the elements of the set assigned to a node are the sets assigned to the children of that node, i.e.

$$d(a) = \{d(x) : a \to x\}.$$

Definition 2.2 The *Anti-Foundation Axiom*, **AFA**, is the statement that every graph has a unique decoration.

3 AFA in constructive set theory

In this section we will present some results about the proof-theoretic strength of systems of constructive set theory with **AFA** instead of $\in$-Induction.

3.1 The theory CZFA

The language of **CZF** is the first order language of Zermelo-Fraenkel set theory, LST, with the non logical primitive symbol $\in$. We assume that LST has also a constant, ω, for the set of the natural numbers.

Definition 3.1 (Axioms of CZF) **CZF** is based on intuitionistic predicate logic with equality. The set theoretic axioms of **CZF** are the following:

1. **Extensionality** $\forall a \, \forall b \, (\forall y \, (y \in a \leftrightarrow y \in b) \rightarrow a = b)$.

2. **Pair** $\forall a \, \forall b \, \exists x \, \forall y \, (y \in x \leftrightarrow y = a \lor y = b)$.

3. **Union** $\forall a \, \exists x \, \forall y \, (y \in x \leftrightarrow \exists z \in a \, y \in z)$.

4. Δ_0 - **Separation scheme** $\forall a \, \exists x \, \forall y \, (y \in x \leftrightarrow y \in a \land \varphi(y))$,

    for every *bounded* formula $\varphi(y)$, where a formula $\varphi(x)$ is bounded, or Δ_0, if all the quantifiers occurring in it are bounded, i.e. of the form $\forall x \in b$ or $\exists x \in b$.

5. **Subset Collection scheme**

    $\forall a \, \forall b \, \exists c \, \forall u \, (\forall x \in a \, \exists y \in b \, \varphi(x, y, u) \rightarrow \exists d \in c \, (\forall x \in a \, \exists y \in d \, \varphi(x, y, u) \land \forall y \in d \, \exists x \in a \, \varphi(x, y, u)))$

    for every formula $\varphi(x, y, u)$.

6. **Strong Collection scheme**

    $\forall a \, (\forall x \in a \, \exists y \, \varphi(x, y) \rightarrow \exists b \, (\forall x \in a \, \exists y \in b \, \varphi(x, y) \land \forall y \in b \, \exists x \in a \, \varphi(x, y)))$

    for every formula $\varphi(x, y)$.

7. **Infinity**

$$(\omega 1) \quad 0 \in \omega \wedge \forall y \, (y \in \omega \to y+1 \in \omega)$$
$$(\omega 2) \quad \forall x \, (0 \in x \wedge \forall y \, (y \in x \to y+1 \in x) \to \omega \subseteq x),$$

where $y+1$ is $y \cup \{y\}$, and 0 is the empty set, defined in the obvious way.

8. **$\in$ - Induction scheme**

$$(IND_\in) \quad \forall a \, (\forall x \in a \, \varphi(x) \to \varphi(a)) \to \forall a \, \varphi(a),$$

for every formula $\varphi(a)$.

Definition 3.2 A mathematically very useful axiom to have in set theory is the *Dependent Choices Axiom*, **DC**, i.e., for all formulae ψ, whenever

$$(\forall x \in a)(\exists y \in a) \, \psi(x,y)$$

and $b_0 \in a$, then there exists a function $f : \omega \to a$ such that $f(0) = b_0$ and

$$(\forall n \in \omega) \, \psi(f(n), f(n+1)).$$

Even more useful in constructive set theory is the *Relativized Dependent Choices Axiom*, **RDC**.[1] It asserts that for arbitrary formulae ϕ and ψ, whenever

$$\forall x \big[\phi(x) \to \exists y \big(\phi(y) \wedge \psi(x,y)\big)\big]$$

and $\phi(b_0)$, then there exists a function f with domain ω such that $f(0) = b_0$ and

$$(\forall n \in \omega) \big[\phi(f(n)) \wedge \psi(f(n), f(n+1))\big].$$

A restricted form of **RDC** is Δ_0-**RDC**: For all Δ_0-formulae θ and ψ, whenever

$$(\forall x \in a)\big[\theta(x) \to (\exists y \in a)(\theta(y) \wedge \psi(x,y))\big]$$

and $b_0 \in a \wedge \phi(b_0)$, then there exists a function $f : \omega \to a$ such that $f(0) = b_0$ and

$$(\forall n \in \omega)\big[\theta(f(n)) \wedge \psi(f(n), f(n+1))\big].$$

Letting $\phi(x)$ stand for $x \in a \wedge \theta(x)$, one sees that Δ_0-**RDC** is a consequence of **RDC**.

---

[1] In Aczel Aczel (1982), **RDC** is called the dependent choices axiom and **DC** is dubbed the axiom of limited dependent choices. We deviate from the notation in Aczel (1982) as it deviates from the usage in classical set theory texts.

Definition 3.3 Let **CZF**$^-$ be the system **CZF** without the $\in$ - Induction scheme and let **CZFA** be the theory **CZF**$^-$ + **AFA**.

CZF$^-$ has certain defects from a mathematical point of view in that this theory appears to be too limited for proving the existence of the functions (as sets of ordered pairs) of addition and multiplication on ω. Likewise, there seems to be no way of proving the existence of the transitive closure of an arbitrary set from the axioms of **CZF**$^-$. The first defect could be cured by just adding axioms which assert the existence of these functions, and this augmentation would then enable one to prove the existence of every primitive recursive functions on ω.[2] However, in this paper I prefer to remedy these defects by slightly strengthening induction on ω to

Σ_1-**IND**$_\omega$ $\phi(0) \wedge (\forall n \in \omega)(\phi(n) \to \phi(n+1)) \to (\forall n \in \omega)\phi(n)$

for all Σ_1 formulae ϕ. It is worth noting that Σ_1-**IND**$_\omega$ actually implies

Σ-**IND**$_\omega$ $\theta(0) \wedge (\forall n \in \omega)(\theta(n) \to \theta(n+1)) \to (\forall n \in \omega)\theta(n)$

for all Σ formulae θ, where the Σ formulae are the smallest collection of formulae comprising the Δ_0 formulae which is closed under $\wedge, \vee$, bounded quantification, and (unbounded) existential quantification. This is due to the fact that every Σ formula is equivalent to a Σ_1 formula provably in **CZF**$^-$. The latter principle is sometimes called the Σ *Reflection Principle* and can be proved like in Kripke-Platek set theory (one easily verifies that the proof of Barwise (1975), I.4.3 also works in **CZF**$^-$). Σ-**IND**$_\omega$ enables one to define functions by Σ recursion on ω (cf. Barwise (1975), I.6) and hence one can prove the existence of every primitive recursive function on ω as well as the transitive closure of an arbitrary set (on the basis of **CZF**$^-$ + Σ_1-**IND**$_\omega$).

We will also study the full scheme of induction on ω,

IND$_\omega$ $\psi(0) \wedge (\forall n \in \omega)(\psi(n) \to \psi(n+1)) \to (\forall n \in \omega)\psi(n)$

for all formulae ψ.

Lemma 3.4 CZF$^-$ + Δ_0-**RDC** *proves* **DC**.

Proof: Assume $(\forall x \in a)(\exists y \in a)\, \psi(x,y)$ and $b_0 \in a$. Then

$$(\forall x \in a)(\exists z)\left[(\exists y \in a)\left(z = \langle x, y \rangle \wedge \psi(x,y)\right)\right].$$

---

[2]The existence of the functions of addition and multiplication on ω can also be proved in **CZF**$^-$ + Δ_0-**RDC**.

Using strong collection there exists a set S such that
$(\forall x \in a)\,(\exists z \in S)\,(\exists y \in a)\,[z = \langle x, y\rangle \wedge \psi(x,y)]$ and

$$(\forall z \in S)\,(\exists x' \in a)\,(\exists y' \in a)\,(z = \langle x', y'\rangle \wedge \psi(x',y')). \tag{1}$$

In particular we have $(\forall x \in a)\,(\exists y \in a)\,\langle x,y\rangle \in S$. Employing Δ_0-**RDC** (with $\phi(x)$ and $\psi(x,y)$ being $x \in a$ and $\langle x,y\rangle \in S$, respectively) there exists a function $f : \omega \to a$ such that $f(0) = b_0$ and $(\forall n \in \omega)\,\langle f(n), f(n+1)\rangle \in S$. By (1) we get $(\forall n \in \omega)\,\psi(f(n), f(n+1))$. □

3.2 Interpreting AFA in Martin-Löf type theory

The constructiveness of **CZF** was shown by Aczel by giving it an interpretation in Martin-Löf's intuitionistic theory of types (cf. Aczel (1978, 1982, 1986)). Lindström (1989) and Hallnäs (1986) have shown that **CZFA** can be interpreted in Martin-Löf type theory as well.

In this subsection we shall recall the interpretation of **CZFA** in Martin-Löf type theory, **MLTT**, as presented in Lindström (1989). In the following we work in **MLTT** with a universe $\mathbb{U}$ closed under the usual type constructors $\Pi, \Sigma, +, I, \mathbb{N}, \mathbb{N}_0, \mathbb{N}_1$ (for details see Martin-Löf (1984)). We will denote the projection functions of the Σ-type by $()_0$ and $()_1$, respectively.

Definition 3.5 (System) A *system* over $\mathbb{U}$ consists of a type S together with an assignment of $\bar{a} \in \mathbb{U}$ and $\tilde{a} \in \bar{a} \to S$ to each $a \in S$.

A system S over $\mathbb{U}$ together with an additional assignment

$$\sup_S(A, f) \in \mathbb{U}$$

to $A \in \mathbb{U}$ and $f \in A \to S$ such that

$$\overline{\sup_S(A, f)} = A \in \mathbb{U} \quad \text{and} \quad (\widetilde{\sup_S(A, f)}) = f \in A \to S$$

will be called a *strong system* over $\mathbb{U}$.

The primordial example of a strong system is Aczel's type of iterative sets V which he used to interpret **CZF** in **MLTT**. Lindström (1989) shows that Martin-Löf type theory allows one to prove that for every system S there exists a maximum bisimulation, $\equiv_S$, on S, and that S equipped with $\equiv_S$ gives rises to an interpretation of **CZFA**.

Definition 3.6 (Bisimulation) A binary relation R on a system S over $\mathbb{U}$ is a *bisimulation* on S if, given $a, b \in S$,

$$R(a,b) \to \forall x \in \bar{a}\, \exists y \in \bar{b}\, R(\tilde{a}x, \tilde{b}y) \wedge \forall y \in \bar{b}\, \exists x \in \bar{a}\, R(\tilde{a}x, \tilde{b}y).$$

The intuitive idea behind the definition of the relation $\equiv_S$ is to define inductively certain approximations $\equiv_n^S$ for each natural number n, and to let $(a \equiv_S b)$ hold if and only if $(a \equiv_n^S b)$ holds for each $n \in \mathbb{N}$ and for $m > n$ the proof of $(a \equiv_m^S b)$ is an extension of the proof of $(a \equiv_n^S b)$.

Definition 3.7 For $a, b \in S$, define $a \equiv_n^S b$ by recursion as follows:

$$\begin{aligned}
(a \equiv_0^S b) &:= \mathbb{N}_1; \\
(a \equiv_{n+1}^S b) &:= (\forall x \in \bar{a})(\exists y \in \bar{b})\,(\tilde{a}x \equiv_n^S \tilde{b}y) \wedge \\
&\quad (\forall y \in \bar{b})(\exists x \in \bar{a})\,(\tilde{a}x \equiv_n^S \tilde{b}y).
\end{aligned}$$

For each natural number n, define projection functions h_n as follows

$$\begin{cases} \mathbf{h}_0 \langle a,b \rangle \langle f,g \rangle = 0_1, & \text{for } \langle f,g \rangle \in (a \equiv_1^S b), \\ \mathbf{h}_{n+1} \langle a,b \rangle \langle f,g \rangle = \langle f',g' \rangle, & \text{for } \langle f,g \rangle \in (a \equiv_{n+2}^S b), \end{cases}$$

where 0_1 is the canonical element of $\mathbb{N}_1$ (the canonical one element set), and

$$f' = (x)\,\langle (fx)_0, \mathbf{h}_n \langle \tilde{a}x, \tilde{b}(fx)_0 \rangle (fx)_1 \rangle,$$
$$g' = (y)\,\langle (gy)_0, \mathbf{h}_n \langle \tilde{a}(gy)_0, \tilde{b}y \rangle (gy)_1 \rangle.$$

Further, let

$$(a \equiv_\infty^S b) = (\Pi n \in \mathbb{N})\,(a \equiv_n^S b).$$

Finally, let $a \equiv_S b$ stand for

$$(\Sigma \chi \in (a \equiv_\infty^S b))\,(\Pi \in \mathbb{N})\, \mathbf{I}\big(a \equiv_n^S b),\, \mathbf{h}_n \langle a,b \rangle \chi(n+1), \chi(n)\big).$$

Proposition 3.8 *The relation $\equiv_S$ is the maximum bisimulation on S. In particular, for $a, b \in S$,*

(i) $(a \equiv_S b) \in \mathbb{U}$.

(ii) $(a \equiv_S b) \to \forall x \in \bar{a}\, \exists y \in \bar{b}\, (\tilde{a}x \equiv_S \tilde{b}y) \wedge \forall y \in \bar{b}\, \exists x \in \bar{a}\, (\tilde{a}x \equiv_S \tilde{b}y)$.

(iii) *If R is a relation on S such that:*

$$R(a,b) \to \forall x \in \bar{a}\, \exists y \in \bar{b}\, R(\tilde{a}x, \tilde{b}y) \wedge \forall y \in \bar{b}\, \exists x \in \bar{a}\, R(\tilde{a}x, \tilde{b}y),$$

then $R(a,b) \to (a \equiv_S b)$.

(iv) $\equiv_S$ *is an equivalence relation on S satisfying*

$$a \equiv_S b \iff \forall x \in \bar{a} \, \exists y \in \bar{b} \, (\tilde{a}x \equiv_S \tilde{b}y) \land \forall y \in \bar{b} \, \exists x \in \bar{a} \, (\tilde{a}x \equiv_S \tilde{b}y).$$

Proof: It can be easily seen that $(a \equiv_S b) \in \mathbb{U}$. See Lindström (1989) for a proof of (ii), (iii), (iv). □

Proposition 3.9 *For every strong system S over a universe $\mathbb{U}$, define the relation $\in_S$ on S by*

$$a \in_S b := (\exists y \in \bar{b})(a \equiv_S \tilde{b}y).$$

Then, as in Aczel's interpretation Aczel (1978), we get an interpretation of the language of set theory in which all theorems of $\mathbf{CZF}^-$ + $\mathbf{IND}_\omega$ + $\mathbf{RDC}$ *are valid.*

Proof: For $\mathbf{CZF}^-$ see Lindström (1989). For $\mathbf{RDC}$ we shall draw on the proof of Aczel (1982), Theorem 5.6. $\alpha \in S$ is said to be *injectively presented* if for all $x, y \in \bar{\alpha}$,

$$\tilde{\alpha}(x) \equiv_S \tilde{\alpha}(y) \to x = y \in \bar{\alpha}.$$

The empty set in the interpretation is witnessed by $\emptyset_S := \sup_S(\mathbb{N}_0, (z)\mathrm{R}_0(z))$. If $\alpha, \beta \in S$ let $\alpha \cup_S \beta \in S$ be defined by $sup_S(\bar{\alpha} + \bar{\beta}, g)$, where g is defined by $g(i(a)) = \tilde{\alpha}(a)$ for $a \in \bar{\alpha}$ and $g(j(a)) = \tilde{\beta}(a)$ for $a \in \bar{\beta}$. For $\alpha \in S$ let $\{\alpha\}_S \in S$ be defined by $\sup_S(\mathbb{N}_1, (z)\mathrm{R}_1(z, \alpha))$. For $\alpha, \beta \in S$ the ordered pair of α and β in the sense of S is defined by $\langle \alpha, \beta \rangle_S := \{\{\alpha\}_S, \{\alpha\}_S \cup_S \{\beta\}_S\}_S$.

The element of S which plays the role of ω in the interpretation is

$$\omega_S = \sup_S(\mathbb{N}, (v)\Delta(v))$$

where $\Delta(0) = \emptyset_S$ and $\Delta(n+1) := \sup_S(\widetilde{\Delta(n)} + \mathbb{N}_1, h)$ with $h(i(a)) = \widetilde{\Delta(n)}(a)$ for $a \in \overline{\Delta(n)}$ and $h(j(a)) = \mathrm{R}_1(a, \Delta(n))$ for $a \in \mathbb{N}_1$.

As $(\forall z \in \mathbb{N}_0)\bot$, it is obvious that $\emptyset_S$ is injectively presented. To show that ω_S is injectively presented we must show that for $n, m \in \mathbb{N}$,

$$\Delta(n) \equiv_S \Delta(m) \to n = m \in \mathbb{N}.$$

This can be shown by a routine double $\mathbb{N}$–induction, first on $n \in \mathbb{N}$ and within that on $m \in \mathbb{N}$.

We are now in a position to prove that $\mathbf{RDC}$ is valid under the interpretation in S. Let B be an $\equiv_S$-extensional species over S, i.e., for all $\alpha, \beta \in S$, whenever $\alpha \equiv_S \beta$, then $B(\alpha) = B(\beta)$. Moreover, let F be an $\equiv_S$-extensional species over $S \times S$ (i.e. $\equiv_S$-extensional in each argument) such that

$$(\forall x \in S)\big(B(x) \to (\exists y \in S)(B(y) \land F(x,y))\big).$$

We claim that for each $\alpha \in S$ such that $B(\alpha)$ there is a $\delta \in S$ such that δ is a function with domain ω_S, $\langle \emptyset_S, \alpha \rangle_S \in_S \delta$ and for every $x \in_S \omega_S$ and all $\beta, \gamma \in S$

$$(\langle x, \beta \rangle_S \in_S \delta \land \langle x', \gamma \rangle_S \in_S \delta) \to \big(B(\beta) \land F(\beta, \gamma)\big).$$

Proof of the claim: Let $\alpha \in S$ such that $B(\alpha)$. Then by the type theoretical version of **RDC** in Aczel (1982), 1.15 there is $c \in \mathbb{N} \to S$ such that $c(0) = \alpha$ and for $n \in \mathbb{N}$

$$B(c(n)) \land F(c(n), c(n+1)). \tag{2}$$

Let $\eta = \sup_S(\mathbb{N}, c)$. Then $\eta \in S$ with $\bar{\eta} = \overline{\omega_S}$. Let

$$\delta := \sup_S(\overline{\omega_S}, (x)\langle \widetilde{\omega_S}(x), \tilde{\eta}(x) \rangle_S).$$

Then δ is a function with domain ω_S as ω_S is injectively presented. Since $\langle \emptyset_S, \alpha \rangle_S \equiv_S \langle \widetilde{\omega_S}(0), \tilde{\eta}(0) \rangle_S \equiv_S \tilde{\delta}(0)$ it follows that $\langle \emptyset_S, \alpha \rangle_S \in_S \delta$. Finally, let $x \in_S \omega_S$ and $\beta, \gamma \in S$ such that $\langle x, \beta \rangle_S \in_S \delta$ and $\langle x', \gamma \rangle_S \in_S \delta$. Then for some $n \in \mathbb{N}$ we have $x \equiv_S \Delta(n)$, so that $x' \equiv_S \Delta(n+1)$ and hence $\beta \equiv_S c(n)$ and $\gamma \equiv_S c(n+1)$. Hence, by (2) and the assumption that B and F are extensional, we get $B(\beta) \land F(\beta, \gamma)$.

If we apply the above result to the extensional predicates defined by formulae in the language of set theory we get the interpretation of **RDC**. $\square$

Systems that satisfy a further completeness property allow for an interpretation of **AFA** as well.

Proposition 3.10 *Given a system S define the type S^* by*

$$S^* := (\Sigma f \in \mathbb{N} \to S)\,(\Pi n \in \mathbb{N})\,\big(f(n) \equiv^S_{n+1} f(n+1)\big).$$

Then S^* is a system under the following assignment of $\overline{(f,g)} \in \mathbb{U}$ and

$$\widetilde{(f,g)} \in \overline{(f,g)} \to S^*,$$

for any $(f,g) \in S^*$:

$$\overline{(f,g)} := \overline{f(0)},$$
$$\widetilde{(f,g)} := (a)\left((n)\, \widetilde{f(n+1)}\, (t_{n+1}a), (n)((g(n+1))_0)_1(t_{n+1}a)\right),$$

where $t_n \in \overline{f(0)} \to \overline{f(n)}$ *for* $n \in \mathbb{N}$ *is given by*

$$t_0 := \mathrm{id}, \qquad t_{n+1} := (a)(((g_n)_0(t_na))_0).$$

Proof: See Lindström (1989), Proposition 3.1. □

Proposition 3.11 *Let S be a strong system. Then the system S^* as defined in Proposition 3.10 can be rendered a strong system under the following assignment* $\sup_{S^*}(A,f) \in \mathbb{U}$ *to* $A \in \mathbb{U}$ *and* $f \in A \to S^*$:

$$\sup_{S^*}(A,f) = (g,h),$$

where $gn \in S$ *for* $n \in \mathbb{N}$ *is given by*

$$f0 = \sup_S(A, (x)((bx)_0 0)), \qquad f(n+1) = \sup_S(A, (x)((bx)_0 n)),$$

and $hn \in (fn \equiv_{n+1} f(n+1))$ *for* $n \in \mathbb{N}$ *is defined by* $hn = (\tau_n, \tau_n)$, *where* $\tau_0 = (x)(x, 0_1)$ *and* $\tau_{n+1} = (x)(x, (bx)_1 n)$.

Proof: See Lindström (1989). □

Theorem 3.12 *Let S be a strong system. Then the interpretation of* **CZF**$^-$ *in the strong system S^* (as defined in Proposition 3.10) also validates* **AFA**.

Proof: See Lindström (1989), Proposition 3.5. □

3.3 Upper bounds

The results of the previous section can be utilized to read off upper bounds for the strengths of the systems **CZFA** + Σ_1-**IND**$_\omega$ and **CZFA** + **IND**$_\omega$ (also in combination with **RDC** and Δ_0-**RDC**). As to the strength of type theory required for the results of the previous subsection, it is pivotal to observe that, unlike Aczel's type of iterative sets, a strong system type is not required to have an inductive structure, i.e. there need not be an elimination rule for it. The strength of a type theory $\mathbf{ML}_1$ with the type constructors $\Pi, \Sigma, +, I, \mathbb{N}, \mathbb{N}_0, \mathbb{N}_1$ and one universe $\mathbb{U}$ closed under these same constructors has been determined by Aczel in Aczel (1977). $\mathbf{ML}_1$ has the same strength as the theory $\Sigma_1^1\text{-}\mathbf{AC}$, a subsystem of second order arithmetic with the Σ_1^1 axiom of choice. Its proof theoretic ordinal is $\varphi\varepsilon_0 0$, where φ denotes the Veblen function (see Schütte (1977)). If one adds a strong system type S to $\mathbf{ML}_1$ the proof theoretic strength does not increase. This can be seen by emulating $\mathbf{ML}_1 + S$ in the theory $\widehat{\mathbf{ID}}_1$ of positive arithmetic fixed points similarly as in Feferman (1982).

Definition 3.13 Let L^+ be the language of Peano arithmetic augmented by a unary predicate symbol Q. For each formula $\phi(Q^+, u)$ of L^+ in which only the variable u occurs free and Q occurs only positively let $\mathbf{I}_\phi$ be a new unary predicate symbol. The *language of* $\widehat{\mathbf{ID}}_1$, $\widehat{\mathbf{L}}$, is the language of Peano arithmetic plus the predicate symbols $\mathbf{I}_\phi$ for each formula $\phi(Q^+, u)$ of L^+.

The axioms of $\widehat{\mathbf{ID}}_1$ comprise the axioms of Peano Arithmetic, with the induction scheme extended to all formulas of $\widehat{\mathbf{L}}$. In addition, $\widehat{\mathbf{ID}}_1$ has the fixed point axioms

$$(\mathbf{FP}) \quad \forall x \left(\phi(\mathbf{I}_\phi, x) \leftrightarrow \mathbf{I}_\phi(x) \right)$$

for all formulae $\phi(Q^+, u)$ of L^+.

Definition 3.14 Occasionally the theory $\widehat{\mathbf{ID}}_1$ is too 'coarse' to obtain exact proof theoretic results. In those situations theories of natural numbers and ordinals which have been introduced by Jäger Jäger (1993) provide a versatile tool. To give an example, the notion of being (a code for) a small type is simulated in $\widehat{\mathbf{ID}}_1$ via a formula which is no longer positive in the fixed point predicates (Aczel's 'trick' in Aczel (1977)) and on account of that the fragment of $\widehat{\mathbf{ID}}_1$ used for the interpretation of $\mathbf{CZFA} + \Sigma_1\text{-}\mathbf{IND}_\omega$ is proof theoretically too strong.

Let $\mathbf{PA}_\Omega^r$ be the theory defined in Jäger (1993). Its main features are that the induction principles on the natural numbers and on the ordinals are restricted to so-called Δ_0^Ω formulae; i.e. formulae in which all the ordinal quantifiers are restricted.

Let $\Sigma^\Omega\text{-}\mathbf{IND}_\mathbb{N}$ be the scheme of induction on the natural numbers for Σ^Ω formulae, i.e. the smallest class of formulae which contains the Δ_0^Ω formulae and is closed under $\wedge, \vee$, quantification over numbers, bounded quantification over ordinals, and (unbounded) existential quantification.

Theorem 3.15 *(i) The type theory used for interpreting the theory* **CZFA** + IND_ω + **RDC** *can be interpreted in* $\widehat{\text{ID}}_1$.

(ii) The type theory used for interpreting the theory **CZFA**+Σ_1-IND_ω +Δ_0-**RDC** *can be interpreted in* $\mathbf{PA}_\Omega^r + \Sigma^\Omega\text{-}\text{IND}_\mathbb{N}$.

Proof: As to (i) one can extend the technique of Aczel (1977) to simultaneously emulate a universe of sets together with a set of iterative sets over it via a positive fixed point definition. All the constructions of section can then be carried out in $\widehat{\text{ID}}_1$.

Ad (ii): In $\mathbf{PA}_\Omega^r$ one can define a universe of codes for types together with a type of iterative sets over it via a positive arithmetical inductive definitions. Thereby the notions of being a code for a type in the universe (a small type) and being a code for an iterative set become Σ^Ω. The notion of being an element of a code for a small type is both Σ^Ω and Π^Ω. Details of the above constructions can be found in Crosilla (2000).

One then has to scrutinize all the type theoretic constructions of section and Lindström (1989) to determine the kind of structural induction and recursion on $\mathbb{N}$ that is required for them. Closer inspection reveals that the strongest form of recursion on $\mathbb{N}$ being used all come in the guise of defining functions from $\mathbb{N}$ to $\mathbb{U}$. Due to the previous observations the interpretation of these constructions can be justified via $\Sigma^\Omega\text{-}\text{IND}_\mathbb{N}$. □

Corollary 3.16 **CZFA**+IND_ω+**RDC** *can be interpreted in* $\Sigma_1^1\text{-}\mathbf{AC}$.

Proof: This follows from the fact that the fixed point axioms of $\widehat{\text{ID}}_1$ can be emulated in $\Sigma_1^1\text{-}\mathbf{AC}$ by interpreting the fixed points as Σ_1^1 sets (cf. Aczel (1977) and Beeson (1985)). □

3.4 Lower bounds

Lower bounds for the theories **CZFA** + Σ_1-IND_ω and **CZFA** + IND_ω can be established by interpreting suitable intuitionistic theories $\mathbf{RA}_\alpha^i$ of the ramified hierarchy up to level α in them. For the definition of the theories $\mathbf{RA}_\alpha^i$ see Feferman and Sieg (1981), chapter II, 1.2. We assume that an ordinal representation system for ε_0 has been formalized in **CZF**$^-$ + Σ_1-IND_ω as a decidable subset of ω (where $A \subseteq \omega$ is said to be decidable if $(\forall n \in \omega)(n \in A \lor n \notin A)$). By Gentzen's proof it follows that $\mathbf{CZF}^- + \mathbf{IND}_\omega$ proves transfinite induction up to α for each (meta) $\alpha < \varepsilon_0$. Using Strong Collection combined with transfinite induction up to α one readily shows the existence of the ramified hierarchy of length α in $\mathbf{CZF}^- + \mathbf{IND}_\omega$. As a result we get that the theory $\mathbf{RA}^i_{<\varepsilon_0} := \bigcup_{\alpha<\varepsilon_0} \mathbf{RA}^i_\alpha$ can be interpreted in $\mathbf{CZF}^- + \mathbf{IND}_\omega$. To show that $\mathbf{RA}^i_{<\omega^\omega}$ can be interpreted in $\mathbf{CZF}^- + \Sigma_1\text{-}\mathbf{IND}_\omega$ we need a preparatory lemma.

Lemma 3.17 *For each (meta)* n, $\mathbf{CZF}^- + \Sigma_1\text{-}\mathbf{IND}_\omega$ *proves*

$$(\forall \alpha)\big[(\forall \beta < \alpha)\phi(\beta) \to \phi(\alpha)\big] \to (\forall \alpha < \omega^n)\phi(\alpha)$$

for every Σ *formula* ϕ. *Here we assume that the variables* $\alpha, \beta, \ldots$ *range over the elements of the ordinal representation system and that* $<$ *denotes the less-relation on them.*

Proof: Recall that $\mathbf{CZF}^- + \Sigma_1\text{-}\mathbf{IND}_\omega$ proves $\Sigma\text{-}\mathbf{IND}_\omega$ as was explained in Definition 3.3, hence we have $\Sigma\text{-}\mathbf{IND}_\omega$ at our disposal. We proceed by meta-induction on n. The case $n = 0$ is trivial. So assume $n = m + 1$ and that the assertion has been shown for m. Suppose

$$(\forall \alpha)\big[(\forall \beta < \alpha)\phi(\beta) \to \phi(\alpha)\big],$$

where ϕ is Σ. Using the induction hypothesis for m we get

$$\forall \gamma \big[(\forall \beta < \gamma)\phi(\beta) \to (\forall \beta < \gamma + \omega^m)\phi(\beta)\big].$$

Using $\Sigma\text{-}\mathbf{IND}_\omega$ with the formula $\psi(x) := (\forall \beta < \alpha + \omega^m \cdot x)\phi(\beta)$ we obtain

$$(\forall \beta < \alpha)\phi(\beta) \to (\forall x \in \omega)(\forall \beta < \alpha + \omega^m \cdot x)\phi(\beta).$$

The latter yields $\forall \alpha \big[(\forall \beta < \alpha)\phi(\beta) \to (\forall \beta < \alpha + \omega^{m+1})\phi(\beta)\big]$ and hence $(\forall \gamma < \omega^n)\phi(\gamma)$. □

Proposition 3.18 (i) $\mathbf{RA}^i_{<\varepsilon_0}$ *can be interpreted in* $\mathbf{CZF}^- + \mathbf{IND}_\omega$.

(ii) $\mathbf{RA}^i_{<\omega^\omega}$ *can be interpreted in* $\mathbf{CZF}^- + \Sigma_1\text{-}\mathbf{IND}_\omega$.

Proof: (i) has been shown. (ii) Using the induction principle from Lemma 3.17 one can show the existence of the ramified hierarchy of sets up to level α for every (meta) $\alpha < \omega^\omega$, and hence one can interpret $\mathbf{RA}^i_{<\alpha}$ in $\mathbf{CZF}^- + \Sigma_1\text{-}\mathbf{IND}_\omega$. □

Theorem 3.19 *(i) The theories* $\mathbf{CZF}^- + \Sigma_1\text{-}\mathbf{IND}_\omega$, $\mathbf{CZFA} + \Sigma_1\text{-}\mathbf{IND}_\omega + \Delta_0\text{-}\mathbf{RDC}$, $\mathbf{CZFA} + \Sigma_1\text{-}\mathbf{IND}_\omega + \mathbf{DC}$, *and* $\Sigma_1^1\text{-}\mathbf{DC}_0$ *are proof-theoretically equivalent. Their proof-theoretic ordinal is* $\varphi\omega 0$.

(ii) The theories $\mathbf{CZF}^- + \mathbf{IND}_\omega$, $\mathbf{CZFA} + \mathbf{IND}_\omega + \mathbf{RDC}$, $\widehat{\mathbf{ID}}_1$, *and* $\Sigma_1^1\text{-}\mathbf{AC}$ *are proof-theoretically equivalent. Their proof-theoretic ordinal is* $\varphi\varepsilon_0 0$.

Proof: It is known that the proof-theoretic ordinal of $\mathbf{RA}^i_{<\omega^\omega}$ is $\varphi\omega 0$; this follows from Theorem 3.2.13 and Theorem 3.1.11 in Feferman and Sieg (1981), chap. I. $\varphi\omega 0$ is the proof-theoretic ordinal of $\Sigma_1^1\text{-}\mathbf{DC}_0$, too, by Cantini (1986). As it can be shown that $\varphi\omega 0$ is also the proof-theoretic ordinal of $\mathbf{PA}_\Omega^r + \Sigma^\Omega\text{-}\mathbf{IND}$, (i) follows from Corollary 3.15 and Proposition 3.18.

The proof that $\varphi\omega 0$ is the proof-theoretic ordinal of $\mathbf{PA}_\Omega^r + \Sigma^\Omega\text{-}\mathbf{IND}$ proceeds as follows. A Gentzen-style (sequent calculus) version of $\mathbf{PA}_\Omega^r + \Sigma^\Omega\text{-}\mathbf{IND}$ allows one to carry out a partial cut elimination in that all cuts with formulae which are neither Σ^Ω nor Π^Ω can be removed. In a next step one shows that such partially normalized derivation of sequents consisting of $\Delta_0^\Omega(\Sigma^\Omega)$ formulae (i.e the smallest class of formulae containing the Σ^Ω formulae which is closed under $\neg, \wedge, \vee$, number quantification, and bounded ordinal quantification) can be interpreted (asymmetrically) in the formal system $\mathbf{RA}_{<\omega^\omega}$ by a method very similar to the one used in Rathjen (1992), Theorem 5.2.

It is also known that the proof-theoretic ordinal of $\mathbf{RA}^i_{<\varepsilon_0}$ is $\varphi\varepsilon_0 0$ (cf. Feferman and Sieg (1981), chap. I., Theorem 3.2.13 and Theorem 3.1.11).

As the latter ordinal is the proof-theoretic ordinal of $\Sigma_1^1\text{-}\mathbf{AC}$ as well, assertion (ii) follows from Corollary 3.16 and Proposition 3.18. □

4 Anti-foundation with inaccessible sets

It has been noted that an important aspect of the contrast between set theory and category theory is that, in the set-theoretic picture, well orderings and the axiom of foundation play a key role, while in the category-theoretic or structuralist picture, well foundedness almost completely disappears. In view of the proof-theoretic results of the previous sections, it is natural to conjecture that constructive category theory can be developed in weak theories.

The main focus of this section will be a constructive set theory in which a great deal of category theory can be formalized à la Grothendieck via universes. The theory is **CZF**$^-$ plus large set axioms which classically are equivalent to large cardinal axioms. Due to the lack of $\in$-induction this system is proof-theoretically weak. On the other hand it is a mathematically rich theory in which one can easily formalize Bishop style constructive mathematics. Crosilla and Rathjen (2000) investigated the strength of **CZF**$^-$ plus the statement that every set is contained in an inaccessible set, **INAC**. Crosilla and Rathjen (2000) showed that **CZF**$^-$ + **INAC** has a realizability interpretation in type theory, thereby establishing that the proof theoretic ordinal of **CZF**$^-$ + **INAC** is a surprisingly small ordinal known as the Feferman-Schütte ordinal Γ_0. The objective of this section is to indicate how the machinery of Crosilla and Rathjen (2000) can be utilized to show that the addition of **AFA** to **CZF**$^-$ + **INAC** does not increase the proof theoretic strength.

The first large set axiom proposed in the context of constructive set theory was the *Regular Extension Axiom*, **REA**, which Aczel introduced to accommodate inductive definitions in **CZF** (cf. Aczel (1978), Aczel (1986)).

Definition 4.1 A set c is said to be *regular* if it is transitive, inhabited (i.e. $\exists u\; u \in c$) and for any $u \in c$ and set $R \subseteq u \times c$ if $\forall x \in u\, \exists y\; \langle x, y\rangle \in R$ then there is a set $v \in c$ such that

$$\forall x \in u\, \exists y \in v\; \langle x, y\rangle \in R \;\wedge\; \forall y \in v\, \exists x \in u\; \langle x, y\rangle \in R.$$

We write $Reg(a)$ for 'a is regular'.

REA is the principle

$$\forall x\, \exists y\; (x \in y \;\wedge\; Reg(y)).$$

Definition 4.2 Let **INAC** be the principle

$$\forall x\, \exists y\; \bigl(x \in y \;\wedge\; Reg(y) \text{ and } y \text{ is a model of } \mathbf{CZF}^-\bigr),$$

i.e. the structure $\langle y, \in\restriction (y \times y)\rangle$ is a model of **CZF**$^-$.

We say that a set is *inaccessible* if it is regular and a model of **CZF**$^-$ and write **INAC**(y) for 'y is inaccessible'.

The formalization of the notion of inaccessibility in Definition 4.2 is somewhat akward as it is very syntactic in that it requires a satisfaction predicate for formulae interpreted over a set. An alternative and more 'algebraic' characterization will be given next.

Definition 4.3 Let $\Omega := \{x : x \subseteq \{0\}\}$. Ω is the class of truth values with 0 representing falsity and $1 = \{0\}$ representing truth. Classically one has $\Omega = \{0,1\}$ but intuitionistically one cannot conclude that those are the only truth values.

For $a \subseteq \Omega$ define

$$\bigwedge a = \{x \in 1 : (\forall u \in a) x \in u\}.$$

A class B is $\bigwedge$-*closed* if for all $a \in B$, whenever $a \subseteq \Omega$, then $\bigwedge a \in B$.

For sets a, b let ${}^a b$ be the class of all functions with domain a and with range contained in b. Let $\mathbf{mv}({}^a b)$ be the class of all sets $r \subseteq a \times b$ satisfying $\forall u \in a\, \exists v \in b\, \langle u, v \rangle \in r$. A set c is said to be *full in* $\mathbf{mv}({}^a b)$ if $c \subseteq \mathbf{mv}({}^a b)$ and

$$\forall r \in \mathbf{mv}({}^a b)\, \exists s \in c\, s \subseteq r.$$

The expression $\mathbf{mv}({}^a b)$ should be read as the collection of *multi-valued functions* from a to b.

Proposition 4.4 ($\mathbf{CZF}^-$) *A set I is inaccessible if and only if the following are satisfied:*

1. *I is a regular set,*

2. *$\omega \in I$,*

3. *$(\forall a \in I) \bigcup a \in I$,*

4. *I is $\bigwedge$-closed,*

5. *$(\forall a, b \in I)\bigl[\{x \in 1 : a = b\} \in I \,\wedge\, \{x \in 1 : a \in b\} \in I\bigr]$.[3]*

6. *$(\forall a, b \in I)(\exists c \in I)\bigl[c \text{ is full in } \mathbf{mv}({}^a b)\bigr]$.*

Proof: See Rathjen (2000b), Proposition 3.4. □

Viewed classically inaccessible sets are closely related to inaccessible cardinals. Let V_α denote the αth level of the von Neumann hierarchy.

Proposition 4.5 (**ZFC**) *A set I is inaccessible if and only if $I = V_\kappa$ for some strongly inaccessible cardinal κ.*

Proof: This is a consequence of the proof of Rathjen and Palmgren (1998), Corollary 2.7. □

---

[3]This clause may be omitted in the presence of $\in$-induction.

Proposition 4.6 *Let* **EM** *denote the principle of excluded middle. The theories* **CZF**$^-$ + **INAC** + **EM** *and*

$$\mathbf{ZFC} + \forall \alpha \, \exists \kappa \, (\alpha < \kappa \, \wedge \, \kappa \text{ is a strongly inaccessible cardinal})$$

have the same proof theoretic strength.

Proof: Crosilla and Rathjen (2000), Lemma 2.10. □

The next two results show that the strength of **INAC** is quite modest when based on constructive set theory.

Theorem 4.7 **CZF** + **REA** *and* **CZF** + **INAC** *have the same proof-theoretic strength as the subsystem of second order arithmetic with* Δ^1_2-*comprehension and bar induction.*

Proof: **CZF** + **INAC** has a realizability in a type theory $\mathbb{MLS}^*$ with the following ingredients:[4]

- $\mathbb{MLS}^*$ demands closure under the usual type constructors $\Pi, \Sigma, +, I, \mathbb{N}, \mathbb{N}_0, \mathbb{N}_1$ (but not the W-type).

- $\mathbb{MLS}^*$ has a superuniverse $\mathbb{S}$ which is closed under $\Pi, \Sigma, +, I, \mathbb{N}, \mathbb{N}_0, \mathbb{N}_1$ and the W-type and the universe operator **U**.

- $\mathbb{MLS}^*$ has a type **V** of iterative sets over $\mathbb{S}$.

- The universe operator **U** takes a type A in $\mathbb{S}$ and a family of types $B : A \to \mathbb{S}$ and produces a universe $\mathbf{U}(A, B)$ in $\mathbb{S}$ which contains A and $B(x)$ for all $x : A$, and is closed under $\Pi, \Sigma, +, I, \mathbb{N}, \mathbb{N}_0, \mathbb{N}_1$ (but not the W-type).

The type **V** then yields a realizability interpretation of **CZF** + **INAC** à la Aczel, using the techniques of Rathjen and Palmgren (1998). On the other hand, $\mathbb{MLS}^*$ can be interpreted in the classical set theory **KPi** by the methods of Rathjen (1994), section 5. As **CZF** + **REA** has the same strength as **KPi** by Rathjen (1994), **CZF** + **REA** and **CZF** + **INAC** also have the same strength. □

Theorem 4.8 *The proof theoretic ordinal of* **CZF**$^-$ + **INAC** *is the Feferman-Schütte ordinal* Γ_0. **CZF**$^-$ + **INAC** *has the same proof theoretic strength as the classical theory* **ATR**$_\mathbf{0}$ *with arithmetical transfinite recursion and induction on the natural numbers restricted to sets.*

---

[4]Regarding the exact formalization of a superuniverse $\mathbb{S}$ and the universe operator **U** see Rathjen (2000b).

Proof: Crosilla and Rathjen (2000), Corollary 9.14. □

The remainder of this section will be devoted to sketching a proof of the following.

Theorem 4.9 *The Feferman-Schütte ordinal Γ_0 is the proof theoretic ordinal of* **CZFA** $+$ **INAC** $+$ Δ_0-**RDC**. **CZFA** $+$ **INAC** $+$ Δ_0-**RDC** *has the same proof theoretic strength as the classical theory* **ATR$_0$**.

4.8 was shown in Crosilla and Rathjen (2000) by giving **CZF$^-$** $+$ **INAC** a realizability interpretation in a theory $\widehat{\mathbf{ID}}^*$ of iterated fixed point definitions. The latter were used to simulate two hierarchies of types, $(\mathbb{U}_\alpha)_\alpha$ and $(\mathbb{V}_\alpha)_\alpha$. $(\mathbb{U}_\alpha)_\alpha$ is a hierarchy of universes and each $\mathbb{V}_\alpha$ is a type of iterative sets over the universes $(\mathbb{U}_\beta)_{\beta<\alpha}$. In particular, $\mathbb{U}_0$ is a universe of small types closed under the usual type constructors (Π, Σ, $+$, I, N, N_k) and $\mathbb{V}_1$ is a type of iterative sets over $\mathbb{U}_0$. New levels in the hierarchy are introduced by reflecting on previous ones, i.e. each new universe contains all the objects of any earlier universe plus a code for that preceding universe. The types of iterative sets $\mathbb{V}_\alpha$ are then endowed with an equivalence relation, $\equiv_\alpha$, which is the maximum bisimulation on $\mathbb{V}_\alpha$, and an elementhood relation, $\in_\alpha$, to allow for a realizability interpretation of **CZF$^-$**.

From now on we assume familiarity with the definitions of Crosilla and Rathjen (2000). Each $\mathbb{V}_\alpha$ gives rise to a strong system over $\mathbb{U}_{<\alpha}$ in the sense of Definition 3.5: If $x \in \mathbb{V}_\alpha$ then $x = \sup(a,b)$ for uniquely determined a, b. Moreover, $a \in \mathbb{U}_{<\alpha}$ and b satisfies $bj \in \mathbb{V}_\alpha$ for every j such that $\mathbb{U}_{<\alpha} \models j \in a$. Therefore we may put $\bar{x} := a$ and $\tilde{x} := b$ to render $\mathbb{V}_\alpha$ a system. Conversely, if $a \in \mathbb{U}_{<\alpha}$ and b satisfies $\mathbb{V}_\alpha \models bj = bi$ for all j, i such that $\mathbb{U}_{<\alpha} \models j = i \in a$, then $\sup(a,b) \in \mathbb{V}_\alpha$, $\overline{\sup(a,b)} = a$, and $\widetilde{\sup(a,b)} = b$.

Definition 4.10 To indicate the dependence on α, we shall denote the relations $\equiv_n$ of Crosilla and Rathjen (2000), Definition 6.3 by $\equiv_n^\alpha$. Guided by Proposition 3.10, we derive a system $\mathbb{V}_\alpha^*$ which enables us to interpret **AFA**:

$$\mathbf{v}_\alpha^* := \sigma\bigl(\pi(\hat{N}, \Lambda x.\hat{\mathbf{v}}_\alpha), \Lambda f.\pi(\hat{N}, \Lambda n.f(n) \equiv_{n+1}^\alpha f(n+1))\bigr),$$
$$x \in \mathbb{V}_\alpha^* \quad \text{iff} \quad \mathbb{U}_{\alpha+1} \models x \in \mathbf{v}_\alpha^*.$$

As shown in Proposition 3.10, $\mathbb{V}_\alpha^*$ can be rendered a strong system. Let $\equiv_\alpha^*$ be the maximum bisimulation on that system.

Note that $\mathbb{U}_{\alpha+1} \models a \equiv_\alpha^* b$ **set** whenever $b, c \in \mathbb{V}_\alpha^*$.

Lemma 4.11 Let $a, b \in \mathbb{V}_\beta^*$ and let $\beta \leq \alpha$. Then $a, b \in \mathbb{V}_\alpha^*$ and

$$\exists x\, \mathbb{U}_{\beta+1} \models x \in (a \equiv_\beta^* b) \;\text{ iff }\; \exists x\, \mathbb{U}_{\alpha+1} \models x \in (a \equiv_\alpha^* b).$$

Proof: The proof is similar to that of Crosilla and Rathjen (2000), Lemma 6.6. □

Definition 4.12 For $a \in \mathbb{V}_\alpha^*$ let $(b \in_\alpha^* a) := \sigma(\bar{a}, \Lambda x. b \equiv_\alpha^* \tilde{a}x)$.

Definition 4.13 Let $x \in \mathbb{V}^*$ stand for $\exists \alpha\, x \in \mathbb{V}_\alpha^*$.
For $a, b \in \mathbb{V}^*$ define

$$(a \equiv_{\mathbb{V}^*} b) := \exists \alpha \left(a \in \mathbb{V}_\alpha^* \wedge b \in \mathbb{V}_\alpha^* \wedge \exists x\, \mathbb{U}_{\alpha+1} \models x \in (a \equiv_\alpha^* b) \right).$$

In addition, for $a \in \mathbb{V}^*$ let $b \in_{\mathbb{V}}^* a := \exists \alpha\, (a \in \mathbb{V}_\alpha^* \wedge b \in_\alpha^* a)$.

Definition 4.14 (Realizability in $\mathbb{V}_\alpha^*$) For each formula $\varphi(x_1, \ldots x_n)$ of **CZF**$^-$ containing at most $x_1, \ldots, x_n$, we define $e \Vdash_\alpha^* \varphi(x_1, \ldots, x_n)$ as follows:

$$
\begin{aligned}
e \Vdash_\alpha^* \bot &:= \mathbb{U}_{\alpha+1} \models e \in \hat{\mathbf{N}}_0; \\
e \Vdash_\alpha^* (x = y) &:= \mathbb{U}_{\alpha+1} \models e \in (x \equiv_\alpha^* y) \wedge \mathbb{V}_\alpha^* \models x, y \text{ set}; \\
e \Vdash_\alpha^* (x \in y) &:= \mathbb{U}_{\alpha+1} \models e \in (x \in_\alpha^* y) \wedge \mathbb{V}_\alpha^* \models x, y \text{ set}; \\
e \Vdash_\alpha^* \psi \wedge \chi &:= e_0 \Vdash_\alpha^* \psi \wedge e_1 \Vdash_\alpha^* \chi; \\
e \Vdash_\alpha^* \psi \vee \chi &:= (e_0 = 0 \rightarrow e_1 \Vdash_\alpha^* \psi) \vee (e_0 \neq 0 \rightarrow e_1 \Vdash_\alpha^* \chi); \\
e \Vdash_\alpha^* \psi \rightarrow \chi &:= \forall q\, (q \Vdash_\alpha^* \psi \rightarrow eq \Vdash_\alpha^* \chi); \\
e \Vdash_\alpha^* \exists x \in a\, \psi(x) &:= \mathbb{U}_{\alpha+1} \models e_0 \in \bar{a} \wedge e_1 \Vdash_\alpha^* \psi(\tilde{a}(e_0)); \\
e \Vdash_\alpha^* \forall x \in a\, \psi(x) &:= \forall i\, (\mathbb{U}_{\alpha+1} \models i \notin \bar{a} \vee ei \Vdash_\alpha^* \psi(\tilde{a}i)); \\
e \Vdash_\alpha^* \exists x\, \psi(x) &:= e_1 \Vdash_\alpha^* \psi(e_0) \wedge \mathbb{V}_\alpha^* \models e_0 \text{ set}; \\
e \Vdash_\alpha^* \forall x\, \psi(x) &:= \forall u \in \mathbb{V}_\alpha^* (eu \Vdash_\alpha^* \psi(u)).
\end{aligned}
$$

We say that a formula φ of **CZF**$^-$ is *realizable in* $\mathbb{V}^*$ if there is an e such that $e \Vdash_\alpha^* \varphi$.

Definition 4.15 (Realizability in $\mathbb{V}^*$) For each formula $\varphi(x_1, \ldots x_n)$ of **CZF** containing at most $x_1, \ldots, x_n$ free, we define $e \Vdash^* \varphi(x_1, \ldots, x_n)$, by replacing in Definition 4.14 $e \Vdash_\alpha^* \ldots$ by $e \Vdash^* \ldots$, and $\mathbb{U}_{\alpha+1}, \mathbb{V}_\alpha^*$ by $\mathbb{U}, \mathbb{V}^*$, respectively, and $\equiv_\alpha^*, \in_\alpha^*$, by $\equiv_{\mathbb{V}^*}$ and $\in_{\mathbb{V}}^*$, respectively.

We say that a formula φ of **CZF** is *realizable in* $\mathbb{V}^*$, or simply realizable, if there is an e such that $e \Vdash^* \varphi$.

Theorem 4.16 *For every α, $\mathbb{V}_\alpha^*$ is a realizability model of* **CZFA** + **RDC**.

Proof: This follows from Proposition 3.9. □

Theorem 4.17 $\mathbb{V}^*$ *is a realizability model of* **CZFA** + **INAC** *as well as* Δ_0-**RDC**. *Moreover, $\mathbb{V}^*$ is also a realizability model of the assertion* $\forall x \exists y \left[x \in y \wedge Reg(y) \wedge y \text{ is a model of } \mathbf{CZFA} \right]$.

Proof: That $\mathbb{V}^*$ realizes **AFA** and Δ_0-**RDC** follows from the fact that all $\mathbb{V}_\alpha^*$ realize **AFA** and **RDC**. That $\mathbb{V}^*$ realizes $\mathbf{CZF}^-$ + **INAC** is proved in the same way as Crosilla and Rathjen (2000), Theorem 8.2 and Theorem 8.4. □

References

Aczel, P. 1977. The strength of Martin–Löf's intuitionistic type theory with one universe. In Miettinen, S. and S. Väänänen, editors, *Proceedings of Symposia in Mathematical Logic, Oulu 1974 and Helsinki 1975*, pages 1–32. University of Helsinki, Department of Philosophy.

Aczel, P. 1978. The type theoretic interpretation of constructive set theory. In MacIntyre, A., L. Pacholski, and J. Paris, editors, *Logic Colloquium '77*. North–Holland, Amsterdam.

Aczel, P. 1982. The type theoretic interpretation of constructive set theory: Choice principles. In Troelstra, A. and D. van Dalen, editors, *The L.E.J. Brouwer Centenary Symposium*. North–Holland, Amsterdam.

Aczel, P. 1986. The type theoretic interpretation of constructive set theory: Inductive definitions. In Marcus, R.B. et al., editors, *Logic, Methodology, and Philosophy of Science VII*. North–Holland, Amsterdam.

Aczel, P. 1988. *Non-well-founded sets*. Number 14 in CSLI Lecture Notes. CSLI Publications, Stanford.

Barwise, J. 1975. *Admissible sets and structures. An approach to definability theory*. Springer Verlag, Berlin.

Barwise, J. and L. Moss. 1996. *Vicious circles*. Number 60 in CSLI Lecture Notes. CSLI Publications, Stanford.

Beeson, M. 1985. *Foundations of Constructive Mathematics*. Springer Verlag, Berlin.

Cantini, A. 1986. On the relation between choice and comprehension principles in second order arithmetic. *Journal of Symbolic Logic*, 51:360–373.

Crosilla, L. 1998. Realizability interpretations for subsystems of **czf** and proof theoretic strength. Technical report, School of Mathematics, University of Leeds.

Crosilla, L. 2000. *Realizability models for constructive set theories wih restricted induction.* PhD thesis, School of Mathematics, University of Leeds.

Crosilla, L. and M. Rathjen. 2002. Inaccessible set axioms may have little consistency strength. *Annals of Pure and Applied Logic*, 114:305–342.

Feferman, S. 1970. Formal theories for transfinite iterations of generalized inductive definitions and some subsystems of analysis. In Myhill, J., A. Kino, and R. Vesley, editors, *Intuitionism and Proof Theory*, pages 303–325. North-Holland, Amsterdam.

Feferman, S. 1982. Iterated inductive fixed-point theories: application to Hancock's conjecture. In *Patras Logic Symposium (Patras, 1980)*, number 109 in Studies in Logic and the Foundations of Mathematics, pages 171–196. North–Holland, Amsterdam.

Feferman, S. and W. Sieg. 1981. Theories of inductive definitions. In Buchholz, W., S. Feferman, W. Pohlers, and W. Sieg, editors, *Iterated inductive definitions and subsystems of analysis*, pages 16–142. Springer, Berlin.

Forti, M. and F. Honsell. 1983. Set theory with free construction principles. *Annali Scuola Normale Superiore di Pisa, Classe di Scienze*, 10:493–522.

Hallnäs, L. 1986. Non wellfounded sets: limits of wellfounded approximations. Technical Report 12, Uppsala University, Department of Mathematics. 16 pages.

Jäger, G. 1993. Fixed points in Peano arithmetic with ordinals. *Annals of Pure and Applied Logic*, 60:119–132.

Lindström, J. 1989. A construction of non-well-founded sets within Martin-Löf type theory. *Journal of Symbolic Logic*, 54:57–64.

Martin-Löf, P. 1984. *Intuitionistic Type Theory*. Bibliopolis, Naples.

Myhill, J. 1975. Constructive set theory. *Journal of Symbolic Logic*, 40: 347–382.

Rathjen, M. 1992. A proof-theoretic characterization of the primitive recursive set functions. *Journal of Symbolic Logic*, 57:954–969.

Rathjen, M. 1994. The strength of some Martin-Löf type theories. *Archive for Mathematical Logic*, 33:347–385.

Rathjen, M. 1998. Realizing Mahlo set theory in type theory. To appear in: Archive for Mathematical Logic.

Rathjen, M. 2000a. The strength of Martin-Löf type theory with a superuniverse. part I. *Archive for Mathematical Logic*, 39:1–39.

Rathjen, M. 2000b. The superjump in Martin-Löf type theory. In Buss, S., P. Hajek, and P. Pudlak, editors, *Logic Colloquium '98*, number 13 in Lecture Notes in Logic, pages 363–386. Association for Symbolic Logic.

Rathjen, M. 2001. Kripke-Platek set theory and the anti-foundation axiom.

Mathematical Logic Quarterly, 47:435–440.

Rathjen, M. and E. Palmgren. 1998. Inaccessibility in constructive set theory and type theory. *Annals of Pure and Applied Logic*, 94:181–200.

Schütte, K. 1977. *Proof theory*. Springer, Berlin.

On Non-wellfounded Constructive Set Theory: Construction of Non-wellfounded Sets in Explicit Mathematics

Sergei Tupailo

> We analyze the proof-theoretic strength of Constructive Set Theory without Foundation, **NCZF**$^-$, with natural numbers as urelements. The upper bounds are established by a realizability interpretation into Explicit Mathematics, which uses the same method as Tupailo (2000). An important feature is building bisimulation between sets "in stages", in a way similar to Lindström (1989). As a corollary we obtain that |**NCZF**$^-$| is bounded by $\varphi(\varepsilon_0, 0)$.

1 Constructive Set Theory with Natural Numbers

The language $\mathcal{L}_{\in,\mathbb{N}}$ of Set Theory with natural numbers as urelements is two-sorted with **number variables** $a, b, c, \ldots$, **set variables** $A, B, C, \ldots$, and two **predicate constants** $=$ and $\in$. $=$ and $\in$ accept arguments of either sort, that is, we are allowed to write $a = b$, $a = B$, $a \in B$, $B \in a$, etc. Additionally we have the following **function constants** which act from numbers to numbers: zero 0, successor $'$, as well as countably many $f_1, f_2, \ldots$ for primitive recursive functions. Number **terms** are built from these in the standard way. The set of free variables of a formula F is denoted by $\mathrm{FV}(F)$, and by $\mathrm{FV}_0(F)$ and $\mathrm{FV}_1(F)$ we denote correspondingly the sets of free number and set variables of F.

---

[*]Address: IAM, University of Bern, Switzerland, sergei@iam.unibe.ch. Research supported by the Swiss National Science Foundation and Estonian Science Foundation. The final version of this paper was prepared during the author's stay at the Institut Mittag-Leffler, Stockholm, Sweden.
I am grateful to the anonymous referee for careful reading of the paper.

Games, Logic, and Constructive Sets
Grigori Mints and Reinhard Muskens (eds.)
Copyright ©2003, CSLI Publications

In free occurrences, we will use o, p, q, r as metavariables for both sorts. In bound occurrences,

$\forall p F[p]$ will denote $\forall X F[X] \wedge \forall x F[x]$, and
$\exists p F[p]$ will denote $\exists X F[X] \vee \exists x F[x]$.

As usual, $\forall X \in A F[X]$ and $\exists X \in A F[X]$ stand for $\forall X (X \in A \to F[X])$ and $\exists X (X \in A \wedge F[X])$, resp., and similarly for $\forall x \in A F[x]$ and $\exists x \in A F[x]$. By F^A we denote the result of replacing each quantifier QX in F by $QX \in A$. *Bounded* (or Δ_0) formulas of $\mathcal{L}_{\in,\mathbb{N}}$ are those built from atoms by means of $\wedge$, $\vee$, $\to$, $\forall x$, $\exists x$, $\forall X \in A$ and $\exists X \in A$.

The logic is *intuitionistic two-sorted with equality*. We take $\bot$ (falsity) as a propositional constant with standard axioms pertaining to it.

The axioms of *Constructive Set Theory with natural numbers* **NCZF** are of three groups: *ontological ones, axioms for natural numbers and axioms for sets*.

Ontological axioms just express the basic features of the set universe with urelements: sets and urelements are different things, and urelements cannot contain anything. These axioms are the following:
$a = A \leftrightarrow \bot$;
$A = a \leftrightarrow \bot$;
$a \in b \leftrightarrow \bot$;
$A \in b \leftrightarrow \bot$.

Arithmetical axioms are the standard axioms of Heyting's arithmetic, where *Induction schema* is taken in the form
$G[0] \wedge \forall x (G[x] \to G[x']) \to \forall x G[x]$, for all formulas $G[x]$.

Set axioms are the following:

Extensionality:
$\forall X \forall Y (\forall p \in X (p \in Y) \wedge \forall p \in Y (p \in X) \to X = Y)$;
Foundation:
$\forall p (\forall q \in p G[q] \to G[p]) \to \forall p G[p]$, for all formulas $G[p]$;
Pair:
$\forall p \forall q \exists Z (p \in Z \wedge q \in Z)$;
Union:
$\forall X \exists Y \forall p \in X \forall q \in p (q \in Y)$;
Infinity:
$\exists X \forall x (x \in X)$;
Δ_0 *Separation:*
$\forall X \exists Y \forall p (p \in Y \leftrightarrow p \in X \wedge F[p])$, for all Δ_0 formulas $F[p]$;

Strong Collection:
$\forall X(\forall p \in X \exists q G[p,q] \rightarrow \exists W(\forall p \in X \exists q \in W G[p,q] \land \forall q \in W \exists p \in X G[p,q]))$, for all formulas $G[p,q]$;

Subset Collection:
$\forall X \forall X' \exists Z(\forall p \in X \exists q \in X' G[p,q] \rightarrow \exists W \in Z(\forall p \in X \exists q \in W G[p,q] \land \forall q \in W \exists p \in X G[p,q]))$, for all formulas $G[p,q]$.

Definition 1 **NCZF**$^-$ *denotes* **NCZF** *without Foundation.*

The theory **NCZF**$^-$ obviously contains Heyting arithmetic, intuitionistic $\mathbf{\Pi}_0^1$–**CA**, and even more. We expect that its strength is bounded below by $\varphi(\varepsilon_0, 0)$.

2 Explicit Mathematics: a reminder

We want to give a realizability interpretation of **NCZF**$^-$ in the system **EETJ** of Explicit Mathematics, which would imply that **NCZF**$^-$ has a pretty low proof-theoretic strength, namely, at most $\varphi(\varepsilon_0, 0)$ (see Feferman (Feferman, 1979, Ch.V, 1)). As was shown in Tupailo (2000), Explicit Mathematics is also an appropriate framework for interpreting Constructive Set Theory (other popular option: Intuitionistic Type Theory, cf. Aczel (1978, 1986), and, most recently, Crosilla (2000)). We have learned in Tupailo (2000) that *Elementary Comprehension* of **EM** suffices for everything, except *Strong Collection* and *Foundation*. In **NCZF**$^-$ we don't have *Foundation*; *Strong Collection* requires, in addition to *Elementary Comprehension*, only the axiom of *Join*.

Before we can get to the job in Section 3, we fix the formulation of Explicit Mathematics we prefer to work with. It's very close to the original formulation of Feferman (1975), but takes some small deviations on a technical level.

Language $\mathcal{L}_{\mathrm{EM}}$. All theories of Explicit Mathematics, considered in this paper, are formulated in a two-sorted language, containing variables for operations (individuals) and names, along with operation constants. Names are thought of as a special kind of operations, coding types (sets) of operations. We use **variables** $a, b, c, \ldots, \mathfrak{a}, \mathfrak{b}, \mathfrak{c}, \ldots$ as ranging over operations, and $\alpha, \beta, \gamma, \ldots$ as ranging over names. The **constants** are the following: combinators k, s, pairing p and projections p_0, p_1, zero 0, successor s_N and predecessor p_N, distinction by cases on natural numbers Δ_0 and join j. As a difference from the canonical system $\mathbf{T_0}$, we omit the inductive generator i. Additionally we have the following 9 **constants** called *name generators*: nat, id, inv, emp, and, or, imp, all, ex. **Terms** are built from variables and constants by the following application clause: if s and t are *terms* then $s \cdot t$ is a *term*, so that the *application* function symbol $\cdot$ accepts arguments of both sorts and returns an operation. **Atomic formulas** are $s = t$ (s coincides with t) and $s \varepsilon t$ (s belongs to the type named by t, s is classified under t), where s and t are terms. **Formulas** are built from atomic formulas by $\wedge, \vee, \rightarrow$ and two kinds of quantifiers, over operations and over names, e.g. $\forall a, \exists a, \forall \alpha, \exists \alpha$. Finally, **expression** is a term or a formula.

Syntactical conventions.
1. We use $e[x]$ for an expression e, possibly containing occurrences of a variable x. In this context by $e[t]$ we mean the result of substituting expression t for all occurrences of x in e.
2. Parentheses in terms are assumed to be associated to the left: e.g., $s \cdot t \cdot u$ is read as $(s \cdot t) \cdot u$.
3. We adopt the following priority among propositional connectives and their abbreviations: $\neg, \wedge, \vee, \rightarrow, \leftrightarrow$. For example, $F_1 \vee \neg F_2 \wedge F_3 \rightarrow F_4 \leftrightarrow F_5$ has to be read as $((F_1 \vee ((\neg F_2) \wedge F_3)) \rightarrow F_4) \leftrightarrow F_5$.

Abbreviations. We use the following shortcuts:
$\neg F \; :\leftrightarrow \; F \rightarrow \bot$;
$F_0 \leftrightarrow F_1 \; :\leftrightarrow \; (F_0 \rightarrow F_1) \wedge (F_1 \rightarrow F_0)$;
$t\downarrow \; :\leftrightarrow \; \exists x(t = x)$;
$\mathcal{N}[t] \; :\Leftrightarrow \; \exists \alpha(t = \alpha)$;
$F[t\downarrow] \; :\Leftrightarrow \; t\downarrow \wedge F[t]$;
$t \doteq \{s[x_1,\ldots,x_n] \,|\, F[x_1,\ldots,x_n]\} \; :\Leftrightarrow \; \mathcal{N}[t] \wedge \forall x(x \varepsilon t \leftrightarrow \exists x_1 \ldots \exists x_n(x = s[x_1,\ldots,x_n] \wedge F[x_1,\ldots,x_n]))$;
$s \simeq t \; :\Leftrightarrow \; (s\downarrow \vee t\downarrow) \rightarrow s = t$;
$s \mathrel{\dot\subseteq} t \; :\Leftrightarrow \; \forall x \varepsilon s(x \varepsilon t)$; $s \mathrel{\dot=} t \; :\Leftrightarrow \; s \mathrel{\dot\subseteq} t \wedge t \mathrel{\dot\subseteq} s$;
$r \mathbin{:} s \mapsto t$ for $\forall x \varepsilon s(rx \varepsilon t)$;
$r \mathbin{:} s^1 \mapsto t$ for $r \mathbin{:} s \mapsto t$, $r \mathbin{:} s^{m+1} \mapsto t$ for $\forall x \varepsilon s(rx \mathbin{:} s^m \mapsto t)$;
$\mathsf{p}_{ij\ldots k}t$ for $\mathsf{p}_k(\ldots (\mathsf{p}_j(\mathsf{p}_i t)) \ldots)$, $i, j, \ldots, k = 0, 0'$;
t' for $\mathsf{s_N} \cdot t$; 1 for $0'$; 2 for $1'$; st for $s \cdot t$; $t(s_1, \ldots, s_n)$ for $(\ldots (ts_1) \ldots s_n)$;
$\langle s, t \rangle$ for $\mathsf{p}st$; $s \neq t$ for $\neg s = t$, etc.

The logic of the theory is *intuitionistic 2-sorted logic of partial terms with equality*. See, e.g., (Beeson, 1985, Ch.VI, 1) or (Troelstra, 1998, 1.3).

The axioms of Explicit Mathematics are divided in several groups. The basic are *applicative axioms, induction on natural numbers, explicit representation, elementary comprehension* **ECA** and *join* **J** (see, e.g., Tupailo (2000)). **EET** is the theory containing all the listed axioms except join, and **EETJ** is the one containing all of them.

EET will be our default theory for reasoning in Explicit Mathematics.

3 Realization of NCZF⁻ into EETJ

For each set variable $A \in \mathcal{L}_{\in,\mathbb{N}}$ we assume a name variable $\alpha_A \in \mathcal{L}_{\text{EM}}$. Sets are interpreted as (names of) trees, whose leaves may be labelled by natural numbers. To begin, we need to set the stage.

We can define a name seq of the type of sequences so that

$$\text{seq} \doteq \{\langle x, y \rangle \mid (x = 0 \wedge y = 0) \vee (x = 1 \wedge y = \langle \mathsf{p}_0 y, \mathsf{p}_1 y \rangle \wedge \mathsf{p}_0 y \, \varepsilon \, \text{seq})\}. \quad (1)$$

To do this, by **ECA** one defines a name seq_0 s.t.

$$\text{seq}_0 \doteq \{\langle 0, 0 \rangle\}, \quad (2)$$

and an operation seq_S s.t.

$$\text{seq}_S \alpha \doteq \{\langle 1, \langle y, z \rangle \rangle \mid y \, \varepsilon \, \alpha\}. \quad (3)$$

Then by primitive recursion one defines an operation sq s.t.

$$\begin{cases} \text{sq } 0 = \text{seq}_0, \\ \text{sq } n' = \text{seq}_S(\text{sq}\, n). \end{cases} \quad (4)$$

Finally by **J** and **ECA** one sets

$$\text{seq} \doteq \{x \mid \exists n \, \varepsilon \, \text{nat}(x \, \varepsilon \, \text{sq}\, n)\}. \quad (5)$$

We abbreviate

$$\text{nil} := \langle 0, 0 \rangle. \quad (6)$$

Now a name lseq of the type of labelled sequences is defined by

$$\text{lseq} := \{\langle 2, \langle y, n \rangle \rangle \mid y \, \varepsilon \, \text{seq} \wedge n \, \varepsilon \, \text{nat}\}. \quad (7)$$

The *label* of a labelled sequence is defined by

$$\text{lb} := \lambda x.\mathsf{p}_1(\mathsf{p}_1 x). \quad (8)$$

In order to work with (labelled) sequences, we define subsidiary operations *length* ln and *concatenation* conc (also written as $*$) by the following equations:

$$\begin{cases} \text{ln nil} \simeq 0, \\ \text{ln}\, \langle 1, \langle a, b \rangle \rangle \simeq \mathsf{s}_\mathsf{N}(\text{ln}\, a), \\ \text{ln}\, \langle 2, \langle a, n \rangle \rangle \simeq \text{ln}\, a; \end{cases} \quad (9)$$

$$\begin{cases} \text{conc}\,(a, \text{nil}) \simeq a, \\ \text{conc}\,(a, \langle 1, \langle c, d \rangle \rangle) \simeq \langle 1, \langle \text{conc}\,(a, c), d \rangle \rangle, \\ \text{conc}\,(a, \langle 2, \langle c, n \rangle \rangle) \simeq \langle 2, \langle \text{conc}\,(a, c), n \rangle \rangle. \end{cases} \quad (10)$$

These operations have expected properties; see (Tupailo, 2000, Prop. 2.1 & 2.2) for details. We note that, in order $\mathsf{conc}\,(s,t)$ to be defined, the first argument s of conc must be a sequence, while the second argument t may be labelled.

Definition 2 ($\sqsupset$)
By **ECA** a name $\sqsupset$ is defined so that

$$\sqsupset \doteq \{\langle x, y\rangle \mid \qquad \qquad \qquad \qquad \qquad \qquad \qquad \quad \tag{11}$$
$$x\,\varepsilon\,\mathsf{seq} \cup \mathsf{lseq} \wedge y\,\varepsilon\,\mathsf{seq} \wedge \exists z\,\varepsilon\,\mathsf{seq} \cup \mathsf{lseq}(z \neq \mathsf{nil} \wedge y*z = x)\}.$$

We will use $x \sqsupset y$, $x \sqsupseteq y$ in place of $\langle x, y\rangle\varepsilon \sqsupset$ and $(x\,\varepsilon\,\mathsf{seq} \cup \mathsf{lseq} \wedge x = y) \vee x \sqsupset y$, resp.

A *set* is a *tree*, i.e. non-empty type of (labelled) sequences downwards closed with respect to $\sqsupset$-relation:

Definition 3 (t names a set, $\mathrm{Set}[t]$)
$\mathrm{Set}[t]$ *is defined as*

$$\mathcal{N}[t] \wedge t \,\dot{\subseteq}\, \mathsf{seq} \cup \mathsf{lseq} \wedge \mathsf{nil}\,\varepsilon\,t \wedge \forall x\,\varepsilon\,t \forall y(x \sqsupset y \to y\,\varepsilon\,t). \tag{12}$$

Note. Sets can be non-wellfounded with respect to $\sqsupset$, i.e. a set may contain an infinite sequence $\ldots \sqsupset x_1 \sqsupset x_0$.

Definition 4 (Subtree operation, str)
By **ECA** we define an operation str in such a way that

$$\mathcal{N}[\mathsf{str}(\alpha, z)] \wedge (x\,\varepsilon\,\mathsf{str}(\alpha, z) \leftrightarrow x\,\varepsilon\,\mathsf{seq} \cup \mathsf{lseq} \wedge z*x\,\varepsilon\,\alpha). \tag{13}$$

Lemma 1 *In* **EETJ** *we have*

$$\mathrm{Set}[\alpha] \wedge z\,\varepsilon\,\alpha \wedge z\,\varepsilon\,\mathsf{seq} \to \mathrm{Set}[\mathsf{str}(\alpha, z)]. \tag{14}$$

Proof. Obvious from the definition. □

As usual, equality between sets is interpreted as *bisimulation*. In order to stay low in proof-theoretic strength, we build bisimulation "in stages", following the way it was done in Lindström (1989). For any sets α and β we want to have an elementary relation $BS[\mathfrak{r}, \alpha, \beta]$ ("$\mathfrak{r}$ is a proof that α and β are bisimulable") and then to check that $BS[\mathfrak{r}, \alpha, \beta]$ has necessary properties. First, by **ECA** we define:

Definition 5 (st, ur)
$$\begin{aligned}\mathsf{st}\,\alpha &\doteq \{x \mid x\,\varepsilon\,\alpha \wedge x\,\varepsilon\,\mathsf{seq} \wedge \mathsf{ln}\,x = 1\}, \\ \mathsf{ur}\,\alpha &\doteq \{x \mid x\,\varepsilon\,\alpha \wedge x\,\varepsilon\,\mathsf{lseq} \wedge \mathsf{ln}\,x = 0\}.\end{aligned} \tag{15}$$

Definition 6 (Bisimulation)
Bisimulation *is a formula* $R[\mathfrak{r}, \alpha, \beta]$ *together with an operation* $\mathfrak{f}$ *such that the following holds:*

if $\mathrm{Set}[\alpha] \wedge \mathrm{Set}[\beta] \wedge R[\mathfrak{r}, \alpha, \beta]$ *then*

$$\forall x \,\varepsilon\, \mathsf{st}\,\alpha \,\big(\mathsf{p}_0(\mathsf{p}_{00}(\mathfrak{fr})x) \,\varepsilon\, \mathsf{st}\,\beta \,\wedge$$
$$R[\mathsf{p}_1(\mathsf{p}_{00}(\mathfrak{fr})x), \mathsf{str}(\alpha, x), \mathsf{str}(\beta, \mathsf{p}_0(\mathsf{p}_{00}(\mathfrak{fr})x))]\big) \wedge$$
$$\forall x\,\varepsilon\,\mathsf{ur}\,\alpha \forall \mathfrak{s}(\mathsf{p}_{01}(\mathfrak{fr})\;\mathsf{lb}(x)\mathfrak{s}\!\downarrow \,\wedge\, x\,\varepsilon\,\mathsf{ur}\,\beta)$$
$$\wedge \qquad\qquad (16)$$
$$\forall y\,\varepsilon\,\mathsf{st}\,\beta \,\big(\mathsf{p}_0(\mathsf{p}_{10}(\mathfrak{fr})y)\,\varepsilon\,\mathsf{st}\,\alpha\,\wedge$$
$$R[\mathsf{p}_1(\mathsf{p}_{10}(\mathfrak{fr})y), \mathsf{str}(\alpha, \mathsf{p}_0(\mathsf{p}_{10}(\mathfrak{fr})y)), \mathsf{str}(\beta, y)]\big) \wedge$$
$$\forall y\,\varepsilon\,\mathsf{ur}\,\beta \forall \mathfrak{s}(\mathsf{p}_{11}(\mathfrak{fr})\;\mathsf{lb}(y)\mathfrak{s}\!\downarrow\,\wedge\, y\,\varepsilon\,\mathsf{ur}\,\alpha).$$

Remark. As long as we consider only bisimulations given by elementary (in parameters) formulas, a bisimulation can be defined as a pair $\langle \mathfrak{n}, \mathfrak{f}\rangle$, where $\mathcal{N}[\mathfrak{n}]$ (meaning: $\mathfrak{n} \doteq \{\mathfrak{r} \mid R[\mathfrak{r}, \alpha, \beta]\}$) and the shown condition holds. This in fact suffices for purposes of the present paper.

Now we are interested to build a maximal bisimulation BS. We set

$$\mathsf{bs}_0 := \lambda a \lambda b.\mathfrak{n}, \qquad (17)$$

where $\mathfrak{n} \doteq \{0\}$. Having in mind that we will have

$$\mathrm{Set}[\alpha] \wedge \mathrm{Set}[\beta] \wedge c\!:\!\mathrm{Set}^2 \mapsto \mathcal{N},$$

where $c\!:\!\mathrm{Set}^2 \mapsto \mathcal{N}$ stands for $\forall \gamma \forall \delta(\mathrm{Set}[\gamma] \wedge \mathrm{Set}[\delta] \to \mathcal{N}[c\gamma\delta])$, we want to define a name $t[\alpha, \beta, c]$ for

$$\{\mathfrak{r}\mid$$
$$\forall x\,\varepsilon\,\mathsf{st}\,\alpha\,\big(\mathsf{p}_0(\mathsf{p}_{00}\mathfrak{r}x)\,\varepsilon\,\mathsf{st}\,\beta\,\wedge$$
$$\mathsf{p}_1(\mathsf{p}_{00}\mathfrak{r}x)\,\varepsilon\,c\cdot\mathsf{str}(\alpha,x)\cdot\mathsf{str}(\beta,\mathsf{p}_0(\mathsf{p}_{00}\mathfrak{r}x))\big) \wedge$$
$$\forall x\,\varepsilon\,\mathsf{ur}\,\alpha\forall\mathfrak{s}(\mathsf{p}_{01}\mathfrak{r}\;\mathsf{lb}(x)\mathfrak{s}\!\downarrow\,\wedge\, x\,\varepsilon\,\mathsf{ur}\,\beta)$$
$$\wedge \qquad\qquad (18)$$
$$\forall y\,\varepsilon\,\mathsf{st}\,\beta\,\big(\mathsf{p}_0(\mathsf{p}_{10}\mathfrak{r}y)\,\varepsilon\,\mathsf{st}\,\alpha\,\wedge$$
$$\mathsf{p}_1(\mathsf{p}_{10}\mathfrak{r}y)\,\varepsilon\,c\cdot\mathsf{str}(\alpha,\mathsf{p}_0(\mathsf{p}_{10}\mathfrak{r}y))\cdot\mathsf{str}(\beta,y)\big) \wedge$$
$$\forall y\,\varepsilon\,\mathsf{ur}\,\beta\forall\mathfrak{s}(\mathsf{p}_{11}\mathfrak{r}\;\mathsf{lb}(y)\mathfrak{s}\!\downarrow\,\wedge\, y\,\varepsilon\,\mathsf{ur}\,\alpha)\}.$$

To do this, first a name $t_1[\alpha, \beta]$ for

$$\{\mathfrak{r}\mid \forall x\,\varepsilon\,\mathsf{st}\,\alpha(\mathsf{p}_0(\mathsf{p}_{00}\mathfrak{r}x)\,\varepsilon\,\mathsf{st}\,\beta)\}$$

is defined by **ECA**. Now we consider a function

$$f := \lambda u.c\cdot\mathsf{str}(\alpha,\mathsf{p}_1 u)\cdot\mathsf{str}(\beta,\mathsf{p}_0(\mathsf{p}_{00}(\mathsf{p}_0 u)(\mathsf{p}_1 u)))$$

and form $t_1' := \mathsf{j}\,(t_1 \times \mathsf{st}\,\alpha, f)$. A name $t_2[\alpha, \beta, c]$ for

$$\{\mathfrak{r} \mid \forall x \, \varepsilon \, \mathsf{st} \, \alpha(\langle\langle \mathfrak{r}, x\rangle, \mathsf{p}_1(\mathsf{p}_{00}\mathfrak{r}x)\rangle \, \varepsilon \, t'_1)\},$$

defined by **ECA**, is then also a name for

$$\{\mathfrak{r} \mid \forall x \, \varepsilon \, \mathsf{st} \, \alpha \, \big(\mathsf{p}_0(\mathsf{p}_{00}\mathfrak{r}x) \, \varepsilon \, \mathsf{st} \, \beta \wedge \\ \mathsf{p}_1(\mathsf{p}_{00}\mathfrak{r}x) \, \varepsilon \, c \cdot \mathsf{str}(\alpha, x) \cdot \mathsf{str}(\beta, \mathsf{p}_0(\mathsf{p}_{00}\mathfrak{r}x)))\}.$$

The second part,

$$\{\mathfrak{r} \mid \forall x \, \varepsilon \, \mathsf{ur} \, \alpha \forall \mathfrak{s}(\mathsf{p}_{01}\mathfrak{r} \, \mathsf{lb}(x)\mathfrak{s}\downarrow \wedge x \, \varepsilon \, \mathsf{ur} \, \beta)\},$$

has a name $t_3[\alpha, \beta]$ by **ECA**.

Symmetrically we construct names t_4, t_5 and t_6 having to do with conjuncts $\forall y \in \beta \ldots$ of 18 and finally build

$$t := \mathsf{and}(\mathsf{and}(t_2, t_3), \mathsf{and}(t_5, t_6)).$$

Now we define an operation bs_S by

$$\mathsf{bs}_\mathsf{S} c = \lambda a \lambda b . t[a, b, c] \tag{19}$$

and formally verify in **EETJ**

$$c : \mathrm{Set}^2 \mapsto \mathcal{N} \to \mathsf{bs}_\mathsf{S} c : \mathrm{Set}^2 \mapsto \mathcal{N}.$$

By primitive recursion we define an operation bs s.t.

$$\begin{cases} \mathsf{bs}\, 0 = \mathsf{bs}_0, \\ \mathsf{bs}\, n' = \lambda a \lambda b . \mathsf{bs}_\mathsf{S}(\,\mathsf{bs}\, n) a b \end{cases} \tag{20}$$

and by induction on **nat** prove

$$\forall n \, \varepsilon \, \mathsf{nat}(\,\mathsf{bs}\, n : \mathrm{Set}^2 \mapsto \mathcal{N}).$$

Now, by primitive recursion, we define projection functions $\mathsf{pr}\, n$ by

$$\begin{cases} \mathsf{pr}\, 0\mathfrak{r} = 0, \\ \mathsf{pr}\, n'\mathfrak{r} = \langle f, g \rangle, \end{cases} \tag{21}$$

where

$$\begin{cases} f := \langle \lambda x . \langle \mathsf{p}_0(\mathsf{p}_{00}\mathfrak{r}x), \mathsf{pr}\, n(\mathsf{p}_1(\mathsf{p}_{00}\mathfrak{r}x))\rangle, \mathsf{p}_{01}\mathfrak{r}\rangle, \\ g := \langle \lambda y . \langle \mathsf{p}_0(\mathsf{p}_{01}\mathfrak{r}y), \mathsf{pr}\, n(\mathsf{p}_1(\mathsf{p}_{01}\mathfrak{r}y))\rangle, \mathsf{p}_{11}\mathfrak{r}\rangle, \end{cases} \tag{22}$$

and by induction on **nat** prove

$$\forall n \, \varepsilon \, \mathsf{nat} \forall \alpha \forall \beta (\mathrm{Set}[\alpha] \wedge \mathrm{Set}[\beta] \to \mathsf{pr}\, n : \mathsf{bs}\, n'\alpha\beta \mapsto \mathsf{bs}\, n\alpha\beta).$$

Definition 7 (Maximal bisimulation $BS[\mathfrak{r}, \alpha, \beta]$)
A formula $BS[\mathfrak{r}, \alpha, \beta]$ is defined as

$$\forall n \, \varepsilon \, \mathsf{nat}(\langle n, \mathfrak{r}n \rangle \, \varepsilon \, \mathsf{j}\,(\mathsf{nat}, \lambda m.\, \mathsf{bs}\, m\alpha\beta) \wedge \mathsf{pr}\, n(\mathfrak{r}n') = \mathfrak{r}n). \quad (23)$$

An $\mathfrak{r}$ such that $BS[\mathfrak{r}, \alpha, \beta]$ is called bisimulator *for α and β and in this case α and β are called* bisimulable by $\mathfrak{r}$.

We **note** that if $\mathsf{Set}[\alpha] \wedge \mathsf{Set}[\beta]$ then $\mathcal{N}[\mathsf{j}\,(\mathsf{nat}, \lambda m.\, \mathsf{bs}\, m\alpha\beta)]$ and $BS[\mathfrak{r}, \alpha, \beta]$ is elementary in $\mathsf{j}\,(\mathsf{nat}, \lambda m.\, \mathsf{bs}\, m\alpha\beta)$. Also, in this case, $BS[\mathfrak{r}, \alpha, \beta]$ is equivalent to

$$\forall n \, \varepsilon \, \mathsf{nat}(\mathfrak{r}n \, \varepsilon \, \mathsf{bs}\, n\alpha\beta \wedge \mathsf{pr}\, n(\mathfrak{r}n') = \mathfrak{r}n).$$

Lemma 2 (*BS* is a bisimulation)
There is an operation ex_0 s.t. $\langle BS, \mathsf{ex}_0 \rangle$ is a bisimulation.

Proof. Assume $\mathsf{Set}[\alpha] \wedge \mathsf{Set}[\beta] \wedge BS[\mathfrak{r}, \alpha, \beta]$.

If $x \, \varepsilon \, \mathsf{ur}\, \alpha$, we simply put $\mathsf{p}_{01}(\mathsf{ex}_0 \mathfrak{r}) := \mathsf{p}_{01}(\mathfrak{r}1)$; symmetrically we put $\mathsf{p}_{11}(\mathsf{ex}_0 \mathfrak{r}) := \mathsf{p}_{11}(\mathfrak{r}1)$. Given $x \, \varepsilon \, \mathsf{st}\, \alpha$, we will find $y \, \varepsilon \, \mathsf{st}\, \beta$ and a bisimulator for $(\mathsf{str}(\alpha, x), \mathsf{str}(\beta, y))$. A similar construction given $y \varepsilon \mathsf{st}\, \beta$ will determine $x \, \varepsilon \, \mathsf{st}\, \alpha$ and a bisimulator for $(\mathsf{str}(\alpha, x), \mathsf{str}(\beta, y))$. So let $x \, \varepsilon \, \mathsf{st}\, \alpha$. For $n \, \varepsilon \, \mathsf{nat}$ let

$$y_n := \mathsf{p}_0(\mathsf{p}_{00}(\mathfrak{r}n')x) \, \varepsilon \, \mathsf{st}\, \beta$$

and

$$\mathsf{t}_n := \mathsf{p}_1(\mathsf{p}_{00}(\mathfrak{r}n')x) \, \varepsilon \, \mathsf{bs}\, n \cdot \mathsf{str}(\alpha, x) \cdot \mathsf{str}(\beta, y_n).$$

Now,
$$\begin{aligned}\mathfrak{r}n' &= \mathsf{pr}\, n'(\mathfrak{r}n''), \\ \mathsf{p}_{00}(\mathfrak{r}n')x &= \mathsf{p}_{00}(\mathsf{pr}\, n'(\mathfrak{r}n''))x = \\ &\langle \mathsf{p}_0(\mathsf{p}_{00}(\mathfrak{r}n'')x),\, \mathsf{pr}\, n(\mathsf{p}_1(\mathsf{p}_{00}(\mathfrak{r}n''))x)) \rangle.\end{aligned} \quad (24)$$

From this it follows that $y_{n'} = y_n$ for all $n \, \varepsilon \, \mathsf{nat}$, and thus $y_n = y_0$ for all $n \varepsilon \, \mathsf{nat}$. So, $\mathsf{t}_n \varepsilon \, \mathsf{bs}\, n \cdot \mathsf{str}(\alpha, x) \cdot \mathsf{str}(\beta, y_0)$. Continuing 24, we have
$$\begin{aligned}\mathsf{p}_1(\mathsf{p}_{00}(\mathfrak{r}n')x) &= \mathsf{pr}\, n(\mathsf{p}_1(\mathsf{p}_{00}(\mathfrak{r}n'')x)), \\ \mathsf{t}_n &= \mathsf{pr}\, n \mathsf{t}_{n'},\end{aligned} \quad (25)$$

so that $BS[\lambda n.\mathsf{t}_n, \mathsf{str}(\alpha, x), \mathsf{str}(\beta, y_0)]$.

Summarizing the construction, we set
$$\mathsf{ex}_0 := \lambda \mathfrak{r}.$$
$$\begin{aligned}&\langle\langle \lambda x.\langle \mathsf{p}_0(\mathsf{p}_{00}(\mathfrak{r}1)x), \lambda n.\mathsf{p}_1(\mathsf{p}_{00}(\mathfrak{r}n')x)\rangle, \mathsf{p}_{01}(\mathfrak{r}1)\rangle, \\ &\langle \lambda y.\langle \mathsf{p}_0(\mathsf{p}_{10}(\mathfrak{r}1)y), \lambda n.\mathsf{p}_1(\mathsf{p}_{10}(\mathfrak{r}n')y)\rangle, \mathsf{p}_{11}(\mathfrak{r}1)\rangle\rangle.\end{aligned} \quad (26)$$

$\square$

Lemma 3 (*BS* is a maximal bisimulation)
There is an operation ex_1 *s.t. if* $\langle R, \mathfrak{f} \rangle$ *is a bisimulation then*

$$\forall \alpha \forall \beta \forall \mathfrak{r}(\mathrm{Set}[\alpha] \wedge \mathrm{Set}[\beta] \wedge R[\mathfrak{r}, \alpha, \beta] \to BS[\mathsf{ex}_1 \mathfrak{fr}, \alpha, \beta]).$$

Proof. Let $\langle R, \mathfrak{f} \rangle$ be a bisimulation and $\mathrm{Set}[\alpha] \wedge \mathrm{Set}[\beta] \wedge R[\mathfrak{r}, \alpha, \beta]$. We let

$$f := \lambda \mathfrak{s}.\mathsf{p}_{00}(\mathfrak{fs}), \quad g := \lambda \mathfrak{s}.\mathsf{p}_{10}(\mathfrak{fs}).$$

Then, by primitive recursion, we define $\mathfrak{t}$ as follows:

$$\begin{cases} \mathfrak{t}0 = \lambda \mathfrak{s}.0, \\ \mathfrak{t}n' = \lambda \mathfrak{s}.\langle\langle \lambda x.\langle \mathsf{p}_0(f\mathfrak{s}x), \mathfrak{t}n\mathsf{p}_1(f\mathfrak{s}x)\rangle, \mathsf{p}_{01}(\mathfrak{fs})\rangle, \\ \qquad\qquad \langle \lambda y.\langle \mathsf{p}_0(g\mathfrak{s}y), \mathfrak{t}n\mathsf{p}_1(g\mathfrak{s}y)\rangle, \mathsf{p}_{11}(\mathfrak{fs})\rangle\rangle. \end{cases}$$

By induction on nat we prove

$$\forall n \; \varepsilon \; \mathsf{nat} \forall \gamma \forall \delta \forall \mathfrak{s}(\mathrm{Set}[\gamma] \wedge \mathrm{Set}[\delta] \wedge R[\mathfrak{s}, \gamma, \delta] \to$$
$$\mathfrak{t}n\mathfrak{s} \; \varepsilon \; \mathsf{bs}\, n\gamma\delta \; \wedge \; \mathsf{pr}\, n(\mathfrak{t}n'\mathfrak{s}) = \mathfrak{t}n\mathfrak{s}).$$

Now, let $\mathsf{ex}_1 := \lambda \mathfrak{f} \lambda \mathfrak{s} \lambda n.\mathfrak{t}[\mathfrak{f}]n\mathfrak{s}$. Then from the previous we have

$$\forall n \; \varepsilon \; \mathsf{nat}(\mathsf{ex}_1 \mathfrak{fr} n \; \varepsilon \; \mathsf{bs}\, n\alpha\beta) \wedge \mathsf{pr}\, n(\mathsf{ex}_1 \mathfrak{fr} n') = \mathsf{ex}_1 \mathfrak{fr} n.$$

Thus, $BS[\mathsf{ex}_1 \mathfrak{fr}, \alpha, \beta]$. $\square$

Lemma 4 *There is an operation* ex_2 *s.t. if* $\mathrm{Set}[\alpha] \wedge \mathrm{Set}[\beta] \wedge BS'[\mathfrak{r}, \alpha, \beta]$, *where* $BS'[\mathfrak{r}, \alpha, \beta]$ *is a formula*

$$\forall x \; \varepsilon \; \mathsf{st}\, \alpha \; \big(\mathsf{p}_0(\mathsf{p}_{00}\mathfrak{r}x) \; \varepsilon \; \mathsf{st}\, \beta \; \wedge$$
$$\qquad BS[\mathsf{p}_1(\mathsf{p}_{00}\mathfrak{r}x), \mathsf{str}(\alpha, x), \mathsf{str}(\beta, \mathsf{p}_0(\mathsf{p}_{00}\mathfrak{r}x))]\big) \bigwedge$$
$$\forall x \; \varepsilon \; \mathsf{ur}\, \alpha \forall \mathfrak{s}(\mathsf{p}_{01}\mathfrak{r} \, \mathsf{lb}(x)\mathfrak{s}{\downarrow} \wedge x \; \varepsilon \; \mathsf{ur}\, \beta) \qquad\qquad (27)$$
$$\bigwedge$$
$$\forall y \; \varepsilon \; \mathsf{st}\, \beta \; \big(\mathsf{p}_0(\mathsf{p}_{10}\mathfrak{r}y) \; \varepsilon \; \mathsf{st}\, \alpha \; \wedge$$
$$\qquad BS[\mathsf{p}_1(\mathsf{p}_{10}\mathfrak{r}y), \mathsf{str}(\alpha, \mathsf{p}_0(\mathsf{p}_{10}\mathfrak{r}y)), \mathsf{str}(\beta, y)]\big) \bigwedge$$
$$\forall y \; \varepsilon \; \mathsf{ur}\, \beta \forall \mathfrak{s}(\mathsf{p}_{11}\mathfrak{r} \, \mathsf{lb}(y)\mathfrak{s}{\downarrow} \wedge y \; \varepsilon \; \mathsf{ur}\, \alpha),$$

then $BS[\mathsf{ex}_2 \mathfrak{r}, \alpha, \beta]$.

Proof. It follows from Lemma 2 that there is an operation ex'_0 s.t. if $\mathrm{Set}[\gamma] \wedge \mathrm{Set}[\delta] \wedge BS'[\mathfrak{s}, \gamma, \delta]$, then

$$\forall x \, \varepsilon \, \mathsf{st} \, \gamma \, \big(\mathsf{p}_0(\mathsf{p}_{00}(\mathsf{ex}'_0\mathsf{s})x) \, \varepsilon \, \mathsf{st} \, \delta \, \wedge$$
$$BS'[\mathsf{p}_1(\mathsf{p}_{00}(\mathsf{ex}'_0\mathsf{s})x), \mathsf{str}(\gamma, x), \mathsf{str}(\delta, \mathsf{p}_0(\mathsf{p}_{00}(\mathsf{ex}'_0\mathsf{s})x))]\big) \bigwedge$$
$$\forall x \, \varepsilon \, \mathsf{ur} \, \gamma \forall \mathsf{s}(\mathsf{p}_{01}\mathsf{r} \, \mathsf{lb}(x)\mathsf{s}{\downarrow} \wedge x \, \varepsilon \, \mathsf{ur} \, \delta)$$
$$\bigwedge \quad (28)$$
$$\forall y \, \varepsilon \, \mathsf{st} \, \delta \, \big(\mathsf{p}_0(\mathsf{p}_{10}(\mathsf{ex}'_0\mathsf{s})y) \, \varepsilon \, \mathsf{st} \, \gamma \, \wedge$$
$$BS'[\mathsf{p}_1(\mathsf{p}_{10}(\mathsf{ex}'_0\mathsf{s})y), \mathsf{str}(\gamma, \mathsf{p}_0(\mathsf{p}_{10}(\mathsf{ex}'_0\mathsf{s})y)), \mathsf{str}(\delta, y)]\big) \bigwedge$$
$$\forall y \, \varepsilon \, \mathsf{ur} \, \delta \forall \mathsf{s}(\mathsf{p}_{11}\mathsf{r} \, \mathsf{lb}(y)\mathsf{s}{\downarrow} \wedge y \, \varepsilon \, \mathsf{ur} \, \gamma),$$

so that $\langle BS', \mathsf{ex}'_0 \rangle$ is a bisimulation. By Lemma 3 we have $BS[\mathsf{ex}_1\mathsf{ex}'_0\mathsf{r}, \alpha, \beta]$, so we set $\mathsf{ex}_2 := \mathsf{ex}_1\mathsf{ex}'_0$. □

Lemma 5 *BS is an equivalence relation, i.e. there are operations* eq_0, eq_1 *and* eq_2 *s.t. if* $\mathrm{Set}[\alpha] \wedge \mathrm{Set}[\beta] \wedge \mathrm{Set}[\gamma]$ *then:*
a) $BS[\mathsf{eq}_0, \alpha, \alpha]$;
b) $BS[\mathsf{r}, \alpha, \beta] \to BS[\lambda n.\mathsf{eq}_1 n\mathsf{r}, \beta, \alpha]$;
c) $BS[\mathsf{r}, \alpha, \beta] \wedge BS[\mathsf{s}, \beta, \gamma] \to BS[\lambda n.\mathsf{eq}_2 n\mathsf{r}\mathsf{s}, \alpha, \gamma]$.

Proof. eq_0, eq_1 and eq_2 are defined by primitive recursion and then by induction on nat necessary properties are proved. Recursive equations are the following:

$$\begin{cases} \mathsf{eq}_0 0 = 0, \\ \mathsf{eq}_0 n' = \langle \langle \lambda x.\langle x, \mathsf{eq}_0 n \rangle, \lambda x \lambda u.0 \rangle, \langle \lambda x.\langle x, \mathsf{eq}_0 n \rangle, \lambda x \lambda u.0 \rangle \rangle; \end{cases}$$

$$\begin{cases} \mathsf{eq}_1 0 = \lambda \mathsf{r}.0, \\ \mathsf{eq}_1 n' = \begin{cases} \lambda \mathsf{r}. \\ \langle \langle \lambda y.\langle \mathsf{p}_0(\mathsf{p}_{10}(\mathsf{r}n')y), \mathsf{eq}_1 n \cdot \mathsf{p}_1(\mathsf{p}_{10}(\mathsf{r}n')y) \rangle, \lambda y \lambda u.0 \rangle, \\ \langle \lambda x.\langle \mathsf{p}_0(\mathsf{p}_{00}(\mathsf{r}n')x), \mathsf{eq}_1 n \cdot \mathsf{p}_1(\mathsf{p}_{00}(\mathsf{r}n')x) \rangle, \lambda x \lambda u.0 \rangle \rangle; \end{cases} \end{cases}$$

$$\begin{cases} \mathsf{eq}_2 0 = \lambda \mathsf{r}\mathsf{s}.0, \\ \mathsf{eq}_2 n' = \begin{cases} \lambda \mathsf{r}\mathsf{s}. \\ \langle \langle \lambda x.\langle \mathsf{p}_0(\mathsf{p}_{00}(\mathsf{s}n')(\mathsf{p}_0(\mathsf{p}_{00}(\mathsf{r}n')x))), \\ \mathsf{eq}_2 n \cdot \mathsf{p}_1(\mathsf{p}_{00}(\mathsf{r}n')x)) \cdot \mathsf{p}_1(\mathsf{p}_{00}(\mathsf{s}n')(\mathsf{p}_0(\mathsf{p}_{00}(\mathsf{r}n')x)))\rangle, \\ \lambda x \lambda u.0 \rangle, \\ \langle \lambda z.\langle \mathsf{p}_0(\mathsf{p}_{10}(\mathsf{r}n')(\mathsf{p}_0(\mathsf{p}_{10}(\mathsf{s}n')z))), \\ \mathsf{eq}_2 n \cdot \mathsf{p}_1(\mathsf{p}_{10}(\mathsf{s}n')z)) \cdot \mathsf{p}_1(\mathsf{p}_{10}(\mathsf{r}n')(\mathsf{p}_0(\mathsf{p}_{10}(\mathsf{s}n')z)))\rangle, \\ \lambda z \lambda u.0 \rangle \rangle. \end{cases} \end{cases}$$

□

Remark. The main distinction of bisimulation defined here from the ones which are often used within set theory setting (see Friedman (1973)), based on the idea of existence of a set $\sim$ s.t.

$$A \sim B \leftrightarrow \forall X \in A \exists Y \in B (X \sim Y) \wedge \forall Y \in B \exists X \in A (X \sim Y),$$

where $U \sim V$ means $\langle U, V \rangle \in \sim$, is that we are actually *building* such a type $\sim$, not merely requiring its *existence*. The advantage is that such a construction requires proof-theoretically only very weak means (namely, in **EM**, only elementary comprehension, join, and induction on natural numbers), while, when standard bisimulation is used, for such an equality interpretation one needs some additional axioms, e.g. Σ_1 *Foundation* (see, for example, (Avigad, 2002, S. 4)), which raise proof-theoretic strength compared to the non-wellfounded version of the theory.

For each function constant $f \in \mathcal{L}_{\in,\mathbb{N}}$ we can define an operation $\mathsf{N}(f)$ representing the same primitive recursive function as f and having the following property: if n is the arity of f then **EET** proves

$$\bigwedge_{i=1}^{n} x_i \, \varepsilon \, \mathsf{nat} \to \mathsf{N}(f) x_1 \ldots x_n \, \varepsilon \, \mathsf{nat}.$$

Now terms of $\mathcal{L}_{\in,\mathbb{N}}$ are translated as follows:

Definition 8 ($\mathsf{N}(t)$)

$$\begin{cases} \mathsf{N}(x) := x; \\ \mathsf{N}(f(t_1, \ldots, t_n)) := \mathsf{N}(f)\mathsf{N}(t_1) \ldots \mathsf{N}(t_n). \end{cases}$$

Definition 9 ($\mathfrak{r}$ realizes F, $\mathfrak{r}$ $\underline{\mathsf{rn}}$ F)

For each formula $F \in \mathcal{L}_{\in,\mathbb{N}}$ we define a formula $(\mathfrak{r} \, \underline{\mathsf{rn}} \, F) \in \mathcal{L}_{\mathrm{EM}}$ with a new free individual variable $\mathfrak{r}$. The definition is given by Table 1.

Remark. According to our notation for substitution, p. 112, in this definition $\mathsf{p}_1 \mathfrak{r} \, \underline{\mathsf{rn}} \, G[\mathsf{p}_0 \mathfrak{r}]$ in the last clause, for example, stands for $(\mathfrak{r} \, \underline{\mathsf{rn}} \, G[X])[\mathfrak{r}/\mathsf{p}_1 \mathfrak{r}][\alpha_X/\mathsf{p}_0 \mathfrak{r}]$.

Definition 10 (Realization, realizable)

1. A term $\mathsf{t} \in \mathcal{L}_{\mathrm{EM}}$ is called realization *of a formula* $F \in \mathcal{L}_{\in,\mathbb{N}}$ *in a theory* $\mathbf{T} \in \mathcal{L}_{\mathrm{EM}}$, *iff*

$$\mathrm{FV}_0(\mathsf{t}) \subseteq \{a \mid a \in \mathrm{FV}_0(F)\} \bigwedge \mathrm{FV}_1(\mathsf{t}) \subseteq \{\alpha_A \mid A \in \mathrm{FV}(F)\}$$

$F \in \mathcal{L}_{\in,\mathbb{N}}$	$(\mathfrak{r} \underline{\mathrm{rn}} F) \in \mathcal{L}_{EM}$
$\bot$	$\bot$
$s = t$	$\mathsf{N}(s) = \mathsf{N}(t)$
$s = A$	$\bot$
$A = s$	$\bot$
$A = B$	$BS[\mathfrak{r}, \alpha_A, \alpha_B]$
$s \in t$	$\bot$
$s \in A$	$\langle 2, \langle \mathsf{nil}, \mathsf{N}(s) \rangle \rangle \; \varepsilon \; \alpha_A$
$A \in s$	$\bot$
$A \in B$	$\mathsf{p}_0\mathfrak{r} \; \varepsilon \; \mathsf{st} \; \alpha_B \wedge BS[\mathsf{p}_1\mathfrak{r}, \alpha_A, \mathsf{str}(\alpha_B, \mathsf{p}_0\mathfrak{r})]$
$F_0 \wedge F_1$	$\mathsf{p}_0\mathfrak{r} \; \underline{\mathrm{rn}} \; F_0 \wedge \mathsf{p}_1\mathfrak{r} \; \underline{\mathrm{rn}} \; F_1$
$F_0 \vee F_1$	$\mathsf{p}_0\mathfrak{r} \; \varepsilon \; \mathsf{nat} \wedge \begin{array}{l}(\mathsf{p}_0\mathfrak{r} = 0 \rightarrow \mathsf{p}_1\mathfrak{r} \; \underline{\mathrm{rn}} \; F_0) \wedge \\ (\mathsf{p}_0\mathfrak{r} \neq 0 \rightarrow \mathsf{p}_1\mathfrak{r} \; \underline{\mathrm{rn}} \; F_1)\end{array}$
$F_0 \rightarrow F_1$	$\forall \mathfrak{x}(\mathfrak{x} \; \underline{\mathrm{rn}} \; F_0 \rightarrow \mathfrak{r}\mathfrak{x}\downarrow \wedge \mathfrak{r}\mathfrak{x} \; \underline{\mathrm{rn}} \; F_1)$
$\forall x G[x]$	$\forall x \; \varepsilon \; \mathsf{nat}(\mathfrak{r}x\downarrow \wedge \mathfrak{r}x \; \underline{\mathrm{rn}} \; G[x])$
$\exists x G[x]$	$\mathsf{p}_0\mathfrak{r} \; \varepsilon \; \mathsf{nat} \wedge \mathsf{p}_1\mathfrak{r} \; \underline{\mathrm{rn}} \; G[\mathsf{p}_0\mathfrak{r}]$
$\forall X G[X]$	$\forall \alpha (\mathsf{Set}[\alpha] \rightarrow \mathfrak{r}\alpha\downarrow \wedge \mathfrak{r}\alpha \; \underline{\mathrm{rn}} \; G[\alpha])$
$\exists X G[X]$	$\mathsf{Set}[\mathsf{p}_0\mathfrak{r}] \wedge \mathsf{p}_1\mathfrak{r} \; \underline{\mathrm{rn}} \; G[\mathsf{p}_0\mathfrak{r}]$

Table 1: Definition of $(\mathfrak{r} \underline{\mathrm{rn}} F)$

and

$$\mathbf{T} \vdash \left(\bigwedge_{a \in \mathrm{FV}_0(F)} (a \; \varepsilon \; \mathsf{nat}) \wedge \bigwedge_{A \in \mathrm{FV}_1(F)} \mathsf{Set}[\alpha_A] \right) \rightarrow \mathfrak{t} \; \underline{\mathrm{rn}} \; F.$$

2. If there exists such a term $\mathfrak{t}$ then F is called realizable in $\mathbf{T}$. We call a theory T_S realizable in $\mathbf{T}$ iff every theorem of T_S is realizable in $\mathbf{T}$. T_S is realizable iff it's realizable in $\mathbf{EET}$.

Theorem 1 *Each theorem of intuitionistic two-sorted predicate calculus with equality is realizable in* **EETJ**.

Proof is standard except for the case of equality axioms. We have to analyze three types of equalities: between numbers, between a number and a set, and between sets. Realizations of equalities between numbers follow from the law $\mathfrak{r} \underline{\mathrm{rn}}\, (s = t) :\Leftrightarrow \mathsf{N}(s) = \mathsf{N}(t)$ (second line of Definition 9) and equality axioms of logic. In particular, $\mathfrak{r}\underline{\mathrm{rn}}(s = t \wedge s \in A) \equiv \mathsf{N}(s) = \mathsf{N}(t) \wedge \langle 2, \langle \mathsf{nil}, \mathsf{N}(s) \rangle \rangle \varepsilon \alpha_A \Rightarrow \langle 2, \langle \mathsf{nil}, \mathsf{N}(t) \rangle \rangle \varepsilon \alpha_A \equiv \mathfrak{r}\underline{\mathrm{rn}}(t \in A)$. Realizations of equalities of the kind $s = A$ follow from the fact that $\mathfrak{r}\underline{\mathrm{rn}}(s = A)$ is $\bot$, and the logical axiom "$\bot \to$ anything". Finally, when considering equalities between sets, the main cases are:

(Eq1) $A = B \wedge A \in C \to B \in C$;
(Eq2) $A = B \wedge C \in A \to C \in B$.

For (Eq1), if $\mathfrak{r}\underline{\mathrm{rn}}\,\alpha = \beta$ and $\mathfrak{s}\underline{\mathrm{rn}}\,\alpha \in \gamma$, then $\mathfrak{s}$ gives us an appropriate point $\mathsf{p_0}\mathfrak{s}$ in γ, and a bisimulator between β and $\mathsf{str}(\gamma, \mathsf{p_0}\mathfrak{s})$ is obtained by commutativity and transitivity (Lemma 5).

For (Eq2), if $\mathfrak{r}\,\underline{\mathrm{rn}}\,\alpha = \beta$ and $\mathfrak{s}\,\underline{\mathrm{rn}}\,\gamma \in \alpha$, then $\mathsf{p_0}\mathfrak{s}\,\varepsilon\,\mathsf{st}\,\alpha \wedge BS[\mathsf{p_1}\mathfrak{s}, \gamma, \mathsf{str}(\alpha, \mathsf{p_0}\mathfrak{s})]$. By $\mathfrak{r}$ and Lemma 2 we obtain $t\,\varepsilon\,\mathsf{st}\,\beta \wedge BS[\mathfrak{q}, \mathsf{str}(\alpha, \mathsf{p_0}\mathfrak{s}), \mathsf{str}(\beta, t)]$ for some t and $\mathfrak{q}$ and by transitivity follows the assertion. □

Now we turn to realizing mathematical axioms of **NCZF⁻** in **EETJ**. According to Theorem 1, this is sufficient to claim realizability of every theorem.

We have to proceed by different groups of axioms of **NCZF⁻**. *Ontological* ones are obviously realized by $\langle \lambda x.x, \lambda y.y \rangle$, since $\mathfrak{r}\,\underline{\mathrm{rn}}\,F \Leftrightarrow \mathfrak{r}\,\underline{\mathrm{rn}}\,\bot \Leftrightarrow \bot$, for F of any of the forms $a = A$, $A = a$, $a \in b$, or $A \in a$. *Arithmetical* axioms are realized as in (Troelstra and van Dalen, 1988, Ch.IV, Section 4). We note that realizing *Induction* requires induction on natural numbers of Explicit Mathematics.

What remains are proofs of the *set* axioms. Those are again realized exactly as in (Tupailo, 2000, S. 3); the only thing to note is that since Δ_0 formulas now may contain quantifiers over urelements, we have to make sure that we still can provide *elementary realization* $\underline{\mathrm{rn}}^{\mathcal{E}}$ for those formulas. But this is indeed so, since the clauses for $\mathfrak{r}\,\underline{\mathrm{rn}}\,\forall x G[x]$ and $\mathfrak{r}\,\underline{\mathrm{rn}}\,\exists x G[x]$ in the Definition 9 are elementary in $\mathfrak{r}\,\underline{\mathrm{rn}}\,G[x]$:

Definition 11 $\mathfrak{r}$ elementarily realizes F, $\mathfrak{r}\,\underline{\mathrm{rn}}^{\mathcal{E}}\,F$

For each bounded formula $F \in \mathcal{L}_{\in,\mathbb{N}}$ we define an elementary formula $(\mathfrak{r}\,\underline{\mathrm{rn}}^{\mathcal{E}}\,F) \in \mathcal{L}_{\mathrm{EM}}$ with a new free individual variable $\mathfrak{r}$. The definition is obtained from the Definition 9 of $\underline{\mathrm{rn}}$ by replacing $\underline{\mathrm{rn}}$ by $\underline{\mathrm{rn}}^{\mathcal{E}}$ everywhere and taking clauses for $\forall X \in A\, G[X]$ and $\exists X \in A\, G[X]$ as follows:

F	$\mathfrak{r}\,\underline{\mathrm{rn}}^{\mathcal{E}} F$
$\forall X \in AG[X]$	$\forall x\,\varepsilon\,\mathsf{st}\,\alpha_A(\mathfrak{r}x{\downarrow} \wedge \mathfrak{r}x\,\underline{\mathrm{rn}}^{\mathcal{E}} G[\mathsf{str}(\alpha_A, x)])$
$\exists X \in AG[X]$	$\mathsf{p}_0\mathfrak{r}\,\varepsilon\,\mathsf{st}\,\alpha_A \wedge \mathsf{p}_1\mathfrak{r}\,\underline{\mathrm{rn}}^{\mathcal{E}} G[\mathsf{str}(\alpha_A, \mathsf{p}_0\mathfrak{r})]$

One establishes correspondence between realizations $\underline{\mathrm{rn}}$ and $\underline{\mathrm{rn}}^{\mathcal{E}}$ in the same way as in Definition 2.11 and Lemma 2.5 of Tupailo (2000):

Lemma 6 Δ_0-lemma, cf. (Tupailo, 2000, L. 2.5)
If $F \in \mathcal{L}_{\in,\mathbb{N}}$ is bounded then there is a term $\mathsf{eq}_F^{\mathcal{E}} \in \mathcal{L}_{\mathrm{EM}}$ s.t. $\mathrm{FV}(\mathsf{eq}_F^{\mathcal{E}}) \subseteq \{a \mid a \in \mathrm{FV}_0(F)\} \bigcup \{\alpha_A \mid A \in \mathrm{FV}_1(F)\}$, $\mathsf{eq}_F^{\mathcal{E}} = \langle \mathsf{p}_0\mathsf{eq}_F^{\mathcal{E}}, \mathsf{p}_1\mathsf{eq}_F^{\mathcal{E}} \rangle$ and the following holds:

$$\mathfrak{x}\,\underline{\mathrm{rn}}\,F \to \mathsf{p}_0\mathsf{eq}_F^{\mathcal{E}}\mathfrak{x}\,\underline{\mathrm{rn}}^{\mathcal{E}} F \tag{29}$$

and

$$\mathfrak{y}\,\underline{\mathrm{rn}}^{\mathcal{E}} F \to \mathsf{p}_1\mathsf{eq}_F^{\mathcal{E}}\mathfrak{y}\,\underline{\mathrm{rn}}\,F. \tag{30}$$

Extensionality is the main axiom for which bisimulation is responsible. However, everything is already contained in Lemma 4.

Lemma 7 (*Extensionality*)
The Extensionality axiom is realizable in **EETJ**.

Proof. Indeed, given $\mathrm{Set}[\alpha] \wedge \mathrm{Set}[\beta]$,

$$\mathfrak{r}\,\underline{\mathrm{rn}}^{\mathcal{E}} \big((\forall X \in \alpha(X \in \beta) \wedge \forall x \in \alpha(x \in \beta)) \wedge$$
$$(\forall Y \in \beta(Y \in \alpha) \wedge \forall y \in \beta(y \in \alpha))\big)$$

reads as

$$\begin{aligned}
\forall x\,\varepsilon\,\mathsf{st}\,\alpha\,\big(\mathsf{p}_0(\mathsf{p}_{00}\mathfrak{r}x)\,\varepsilon\,\mathsf{st}\,\beta\,\wedge& \\
BS[\mathsf{p}_1(\mathsf{p}_{00}\mathfrak{r}x), \mathsf{str}(\alpha, x), \mathsf{str}(\beta, \mathsf{p}_0(\mathsf{p}_{00}\mathfrak{r}x))]\big) &\bigwedge \\
\mathsf{p}_{01}\mathfrak{r}\,\underline{\mathrm{rn}}^{\mathcal{E}}\forall x \in \alpha(x \in \beta)& \\
\bigwedge& \\
\forall y\,\varepsilon\,\mathsf{st}\,\beta\,\big(\mathsf{p}_0(\mathsf{p}_{10}\mathfrak{r}y)\,\varepsilon\,\mathsf{st}\,\alpha\,\wedge& \\
BS[\mathsf{p}_1(\mathsf{p}_{10}\mathfrak{r}y), \mathsf{str}(\alpha, \mathsf{p}_0(\mathsf{p}_{10}\mathfrak{r}y)), \mathsf{str}(\beta, y)]\big) &\bigwedge \\
\mathsf{p}_{11}\mathfrak{r}\,\underline{\mathrm{rn}}^{\mathcal{E}}\forall y \in \beta(y \in \alpha),&
\end{aligned} \tag{31}$$

$$\begin{gathered}
\mathsf{p}_{01}\mathsf{r}\ \underline{\mathsf{rn}}^{\mathcal{E}}\forall x{\in}\alpha(x\in\beta) \equiv \\
\forall x\ \varepsilon\ \mathsf{nat}\forall \mathsf{s}(\mathsf{s}\ \underline{\mathsf{rn}}\ (x\in\alpha) \to \mathsf{p}_{01}\mathsf{r}x\mathsf{s}{\downarrow}\ \underline{\mathsf{rn}}\ (x\in\beta)) \equiv \\
\forall x\ \varepsilon\ \mathsf{ur}\ \alpha\forall \mathsf{s}(\mathsf{p}_{01}\mathsf{r}\ \mathsf{lb}(x)\mathsf{s}{\downarrow}\ \wedge\ x\ \varepsilon\ \mathsf{ur}\ \beta), \\
\mathsf{p}_{11}\mathsf{r}\ \underline{\mathsf{rn}}^{\mathcal{E}}\forall y{\in}\beta(y\in\alpha) \equiv \\
\forall y\ \varepsilon\ \mathsf{nat}\forall \mathsf{s}(\mathsf{s}\ \underline{\mathsf{rn}}\ (y\in\beta) \to \mathsf{p}_{11}\mathsf{r}y\mathsf{s}{\downarrow}\ \underline{\mathsf{rn}}\ (y\in\alpha)) \equiv \\
\forall y\ \varepsilon\ \mathsf{ur}\ \beta\forall \mathsf{s}(\mathsf{p}_{11}\mathsf{r}\ \mathsf{lb}(y)\mathsf{s}{\downarrow}\ \wedge\ y\ \varepsilon\ \mathsf{ur}\ \alpha),
\end{gathered} \tag{32}$$

and then a bisimulator between α and β is obtained by Lemma 4. □

The only axiom which requires a new proof is the axiom of *Infinity*. This is given below.

Lemma 8 (*Infinity*)
Axiom Infinity is realizable in **EETJ**.

Proof. An infinite tree is constructed as
$$\{x \mid x = \mathsf{nil} \vee \exists n\ \varepsilon\ \mathsf{nat}(x = \langle 2, \langle \mathsf{nil}, n\rangle\rangle)\}. \tag{33}$$
□

These facts give us

Theorem 2 **NCZF**$^-$ *is realizable in* **EETJ**; *therefore its strength is bounded above by* $\varphi(\varepsilon_0, 0)$.

Proof. Realizability has been shown. The fact about $\varphi(\varepsilon_0, 0)$ follows from $|\mathbf{EETJ}| = \varphi(\varepsilon_0, 0)$ (Feferman, 1979, Ch.V, 1). □

References

Aczel, P. 1978. The type theoretic interpretation of constructive set theory. In A. MacIntyre, L. Pacholski, J. P., editor, *Logic Colloquium '77*, pages 55–66.

Aczel, P. 1986. The type theoretic interpretation of constructive set theory: inductive definitions. In et al., R. M., editor, *Logic, Methodology and Philosophy of Science VII*.

Avigad, J. 2002. Interpreting classical theories in constructive ones. *Journal of Symbolic Logic*. To Appear.

Barwise, J. 1975. *Admissible sets and structures*. Springer.

Beeson, M. 1985. *Foundations of Constructive Mathematics*. Springer.

Crosilla, L. 2000. *Realizability models for constructive set theories with restricted induction*. PhD thesis, University of Leeds.

Feferman, S. 1975. A language and axioms for explicit mathematics. In *Algebra and Logic*, volume 450 of *Lecture Notes in Mathematics*, pages 87–139. Springer, Berlin.

Feferman, S. 1979. Constructive theories of functions and classes. In *Logic Colloquium '78*, pages 159–224.

Friedman, H. 1973. The consistency of classical set theory relative to a set theory with intuitionistic logic. *Journal of Symbolic Logic*, pages 315–319.

Lindström, I. 1989. A construction of non-well-founded sets within Martin-Löf's type theory. *Journal of Symbolic Logic*, 54(1).

Troelstra, A. 1998. Realizability. In Buss, S., editor, *Handbook of Proof Theory*, pages 407–474. North Holland.

Troelstra, A. and D. van Dalen. 1988. *Constructivism in Mathematics*, volume I–II. North Holland.

Tupailo, S. 2000. Realization of Constructive Set Theory into Explicit Mathematics: a lower bound for impredicative Mahlo universe. Technical Report IAM-00-004, University of Bern, Switzerland. To appear in the Annals of Pure and Applied Logic.

Tupailo, S. 2001. Realization of analysis into Explicit Mathematics. *Journal of Symbolic Logic*, 66(4): 1848–1864.

Index

Absent-minded driver game, 29
AFA, *see* Anti-Foundation Axiom
affine logic, 61
antecedent, 75
Anti-Foundation Axiom, 87–106

Backward induction, 19
binary formula, 68
bisimulation, 7, 9–11, 13, 43, 45, 50, 93, 109, 114, 115, 117–120, 123
    class, 47, 55
    maximum, 92, 93, 104, 115, 117, 118
    modal bisimulation, 12, 13
    power bisimulation, 11, 12
bounded (Δ_0) formula, 110

Centipede, 5
consistency condition, 28, 29
constructive set theory, 87–124
Constructive Zermelo-Fraenkel set theory, 88
cut free proof, 80
cut inference, 80
cut-elimination, 75–83
CZF, *see* Constructive Zermelo-Fraenkel set theory

Deal of cards, 49
Dependent Choices Axiom, 90

Explicit Mathematics, 109, 111
Extensionality Axiom, 89, 110, 123

extensive games
    of imperfect information, 5, 6, 16, 25, 26
    of perfect information, 5, 16, 24

Factual knowledge, 50, 51
Feferman-Schütte ordinal, 103, 104
fixed point, 55
Foundation Axiom, 110

Game algebra, 13, 14, 21
game equivalence, 13, 20
game operations, 13
game semantics, 61, 73
game state, 43–57
game-validity, 74

Herbrand form, 63, 64, 66–73
Herbrand's theorem, 63
Hexa, 46–49

Immediate majorant, 78
inaccessible cardinal, 102
inaccessible set, 100–102
$\in$ - Induction scheme, 90
Infinity Axiom, 90, 110
information set, 26
informational independence, 24–41
initial game states, 49
interpretability, 98, 99

Kripke frame, 30
Kripke structure, 31

Linear logic, 61
logic games, 3

Majorant, 76
Martin-Löf type theory, 88, 92
monotone formula, 75–78, 80, 82
monotonic formula, *see* monotone formula

Nash equilibrium, 5
non-wellfounded set, 87, 109–124

Pair Axiom, 89, 110
password game, 25, 27, 39
perfect recall, 30
pre-initial state, 54
preference logics, 6
Prisoner's Dilemma, 4
private ignorance, 50, 52
private knowledge, 50, 52
process equivalence, 13
process theories, 7
public knowledge, 50

Realizability, 105, 109, 111, 122
realizability model, 106
realizable, 120, 121, 123, 124
realization, 120
    elementary, 122
Regular Extension Axiom, 101
regular set, 101
Relativized Dependent Choices Axiom, 90
resource conscious, 61–74

Semantical game, 33
    of imperfect information, 35
Δ_0 Separation scheme, 89, 110
signalling, 27
simple Herbrand validity, 61, 67, 68, 72, 74

Stag Hunt, 4
Strong Collection scheme, 89, 111
strong system, 92, 94, 96, 97, 104
strongly inaccessible cardinal, 102
Subset Collection scheme, 89, 111
succedent, 75
syntactic relativization, 56

Uniform strategy, 17
uniformity, 26, 27
Union Axiom, 89, 110
universal Horn formula, 76, 78, 80
universal simple Herbrand validity, 68, 72, 74

Von Neumann & Morgenstern condition, 29